Green Energy and Technology

Climate change, environmental impact and the limited natural resources urge scientific research and novel technical solutions. The monograph series Green Energy and Technology serves as a publishing platform for scientific and technological approaches to "green"—i.e. environmentally friendly and sustainable—technologies. While a focus lies on energy and power supply, it also covers "green" solutions in industrial engineering and engineering design. Green Energy and Technology addresses researchers, advanced students, technical consultants as well as decision makers in industries and politics. Hence, the level of presentation spans from instructional to highly technical.

Indexed in Scopus.

Indexed in Ei Compendex.

Carlos Eduardo Keutenedjian Mady

Exergy Analysis

From Cells to Societies: A Unified Approach

Springer

Carlos Eduardo Keutenedjian Mady
São Paulo, Brazil

ISSN 1865-3529 ISSN 1865-3537 (electronic)
Green Energy and Technology
ISBN 978-3-032-01645-4 ISBN 978-3-032-01646-1 (eBook)
https://doi.org/10.1007/978-3-032-01646-1

This Springer imprint is published by the registered company Springer Nature Switzerland AG
The registered company address is: Gewerbestrasse 11, 6330 Cham, Switzerland

Using The Cramberries words: this is an "ode to my family".

I dedicate this to my family: Melissa, Eduardo, Maria Carolina and Charles Mady. You are my real inspiration and exergy for always keeping forward.

Preface

This book aims to spread the application of exergy analysis without compromising the technical rigor required in this field of knowledge. The use of exergy analysis (second law of thermodynamics) helps us understand the complexities of the correlation between our daily exergy expenditure to survive, the exergy necessary to maintain a minimal sustainable life, and the consequences of not changing our mindsets regarding the rational use of natural resources.

The first chapters provide an initial introduction to the conceptualization of the term exergy and the advantages of assessing biological phenomena using thermodynamics, including the maintenance of life on Earth with minimal irreversibilities. All living beings are in a state of disequilibrium with their environment. We experience higher temperatures (as homeothermic beings), higher pressures (as measured by our doctor during each consultation), and a different composition (or chemical potential) than the environment. Some compounds exist only inside the body and are a mixture of organic and inorganic ones, such as ATP and hemoglobin.

The macroscopic changes in our thermodynamic behavior, such as thermal sensation, can be analyzed from the perspective of energy conversion quality, aiming to decrease air conditioning consumption. Or even to understand how we expend "a lot of exergy" to have a simple meal with low calories since there is a production chain and irreversibilities associated with each step of production, transportation, and so on.

Moreover, is it possible to create sports indicators and compare them with the ones proposed in medical studies of sports? How does the exergy behavior of the human body change under physical activity? What happens when we achieve hyperthermia? Regarding medical care, how to induce hypothermia in patients after a cardiorespiratory arrest? How can a pregnant woman in an urban environment understand if the pollution can affect her fetus? These questions were raised by me and other researchers who made this book possible, so I must acknowledge some of them, starting with Prof. Silvio de Oliveira Junior. He is the primary expert in our group in this area and continues to captivate undergraduates and graduate students with his perspective on exergy and profound knowledge of this field.

The other researchers referenced and not referenced over the text: (1) Dr. Rafael Fernandes Mosquim (2) Dr. Laryssa Abdala (3) M.Sc. Juliana Rangel Cenzi (4) M.Sc. Clara Reis Pinto (5) M.Sc. Ellen Cavalcante Alves (6) M.Sc. Lucas Sandoli Lima (7) M.Sc. Débora Silva Molliet (8) M.Sc. Marina Torelli Reis Martins Pereira (9) M.Sc. Felipe Godoy Righetto (10) M.Sc. Henrique Naim Finianos Feliciano (11) Dr. Fernando Fusco Rovai (12) Eng. Ingrid Rossilho Casale (13) M.Sc Felipe Godoy Righetto (14) Prof. Dr. Izabela Batista Henriques (15) Prof. Dr. Cyro Albuquerque.

This work is part of a project that began with the first steps of my undergraduate course in Mechanical Engineering, and with my Ph.D., I have been trying to understand the relationships between thermodynamics and biological phenomena for quite some time; each step we take opens more doors than it closes.

Finally, I must use the sentence of Prof. Oliveira-Junior: "Our main challenge is to live better, more, and using less Exergy!"

São Paulo, Brazil Carlos Eduardo Keutenedjian Mady

Competing Interests The author has no competing interests to declare that are relevant to the content of this manuscript.

The author thanks the National Council for Scientific and Technological Development (CNPq, productivity grants 307405/2021-4, 304395/2018-8) and UNIVERSAL project 400401/2016-9. The author also acknowledge FAPESP (São Paulo research foundation) for project 2015/22883-7. Several scholarships made this book viable from CAPES, CNPq and FAPESP.

Ethics Approval In Chap. 4, all tests were conducted following approved protocols at the Faculty of Medicine of the University of São Paulo, with ethical clearance from the Ethics Committee for Analysis of Research Projects (CAPPesq) under registration numbers 16507 and 2477702. It is worth noting that such protocols are designed to induce exhaustion in participants.

Contents

Symbols

B	Exergy (kJ)
$\dot{B}$	Exergy rate (kW)
b	Specific exergy (kJ/kg)
c	Specific heat (kJ/kgK)
c_p	Specific heat at constant pressure (kJ/kgK)
c_v	Specific heat at constant volume (kJ/kgK)
G	Gibbs free energy (kJ)
g	Specific free energy (kJ/kg)
H	Enthalpy (kJ)
$\dot{H}$	Enthalpy rate (kW)
h	Specific enthalpy (kJ/kg)
LHV	Lower heating value (kJ/kg)
m	Mass (kg)
$\dot{m}$	Mass flow rate (kg/s)
n	Number of mols (kmol, mol)
P	Pressure (kPa, bar)
Q	Heat transfer (kJ)
$\dot{Q}$	Heat transfer rate (kW)
$\dot{S}$	Entropy rate (kW/K)
S	Entropy (kJ/K)
s	Specific entropy (kJ/kgK)
T	Temperature (°C, K)
U	Internal energy (kJ)
u	Specific internal energy (kJ/kg)
V	Velocity (m/s)
$\bar{R}$	Universal gas constant (kJ/kmolK)
R_g	Gas constant (kJ/kgK)
$\forall$	Volume (m^3)
$\dot{\forall}$	Volumetric flow rate (m^3)
v	Specific volume (m^3/kg)
W	Work transfer (kJ)

$\dot{W}$	Power transfer (kW)
x	Molar os mass fraction in liquid state (–)
y	Molar os mass fraction in gaseous state (–)
z	Height

Greek Letters

δ	Inexact differential
η	Efficiency (%)
μ	Chemical potential (kJ)
ω	Blood perfusion rate (m^3)
σ	Irreversibilities or "entropy generation" (kJ/K)
$\dot{\sigma}$	Irreversibilities or "entropy generation" rate (kJ/K)

Subscripts

0	Environmental condition
$\Delta\mu$	Chemical potential
ΔT	Thermal exergy
ΔP	Mechanical exergy
a	Air or arterial blood
B	Related to exergy
bl	Blood
ch	Chemical exergy
CV	Control volume
d	Destroyed exergy
dest	Destroyed exergy
in	Input
O_2, max	Maximum oxygen consumption
MAX	Maximum
out	Output
t	Tissue
v	Venous blood

Chapter 1
Introduction

1.1 Introductory Comments

There is no easy way to define the exergy, which refers to the maximum potential to perform work in an energy conversion process (or quality of the energy). The first proposition of this term can refer to Rant [1]. Opposition to acceptance of this concept is common in academic and industrial contexts. It probably originates from the imprecise definition of energy, intrinsic to everyone, which is "a certain capacity to produce a **certain** effect." Nevertheless, Shukuya [2] argues that each type of energy has a different effect—electrical, mechanical, chemical, among others. For instance, 100 kJ of heat transfer does not have the same thermodynamic effect as 100 kJ of mechanical energy. Differently, 100 kJ of exergy is always the same.

The principle of energy conservation, also known as the first law of thermodynamics, states that energy can be converted from one form to another. However, it can never be created or destroyed, whereas such a law does not provide information about its quality. As highlighted by Szergut et al. [3], the second law of thermodynamics expands the discourse by not only setting limits on the types of processes that can occur but also introducing such nuances regarding energy quality, making it a widely used terminology among experts in the field. One can raise the question of: How can we expand this concept to society, medicine, sports, pathology diagnosis, and policymakers? Including this type of analysis in decision-making is a challenge for our generation. Which alternative uses better natural resources and with maximum quality of energy (or exergy efficiency)? Then maybe we can state that "our main challenge is to live better, more, and using less exergy." I understand this book as a logical continuation of [4]

As discussed in the works by [5] and [6], an alternative approach to assessing the decrease in the ability to perform work or the reduction in the quality of energy is through the concept of entropy generation and the application of the Gouy–Stodola theorem–$\dot{W}_{lost} = T_0\dot{\sigma}$ (lost power equals the environmental temperature multiplied by irreversibilities). Such an approach has gained wider acceptance than the

C. E. Keutenedjian Mady, *Exergy Analysis*, Green Energy and Technology,
https://doi.org/10.1007/978-3-032-01646-1_1

straightforward application of the second law of thermodynamics since it gives a clue about process efficiency and provides a physical interpretation of irreversibilities (entropy generation).

Exergy provides a measure of the ability of a substance to perform a specific effect. It quantifies the amount of work that can potentially be extracted from an energy source, in the limit, how much electrical energy one can produce—"a noble form of energy"; other forms of energy/exergy direct conversion are shaft work, potential energy, kinetic energy. With such an approach, it is possible to discuss how far the process is from the ideal. Moreover, it is a method to quantify inefficiencies in energy conversion processes and, therefore, **evaluate how "satisfactory" or "inadequate" certain processes are using natural resources** [4].

Entropy is a non-conservative property, demonstrated by the second law of thermodynamics, where there are always traces of irreversibilities or "entropy generation". It is necessary to remember that these former thermodynamic effects are not properties but a function of the energy conversion process. In the natural process, there is always an increase of the irreversibilities. One of the paths to enunciate the second law of thermodynamics is that for an isolated system, the **property** entropy always increases ($dS_{U,\forall} \geq 0$), characterizing the thermodynamic equilibrium at constant internal energy (U [kJ]) and volume ($\forall$ [m^3]). Hence, each process makes the system approach the equilibrium, maximizing its entropy. Using Gibbs coordinates and relation ($TdS = dU + Pd\forall$), it is possible to obtain Eq. 1.1, where $S = S(U, \forall)$, for a system with a single substance. As it will be seen in Chap. 2, obtaining the thermodynamic statement of equilibrium from this expression is possible. For all changes in the thermodynamic state, the approach of the equilibrium is always towards maximizing the **property** entropy at constant volume ($d\forall = 0$) and internal energy ($dU = 0$)–isolated system.

$$dS(U, \forall) = \left.\frac{\partial S(U, \forall)}{\partial U}\right|_{\forall} dU + \left.\frac{\partial S(U, \forall)}{\partial \forall}\right|_{U} d\forall = \frac{1}{T}dU + -Pd\forall \tag{1.1}$$

Considering Gibbs free energy, using the same coordinates, but $G = G(T, P)$ we can use Eq. 1.2. By fixing the temperature and pressure ($dG_{T,P} \leq 0$), an isolated system achieves its minimum when at the same thermodynamic equilibrium, but at constant temperature (T [K]) and pressure (P [kPa]). The only modification is the thermodynamic property we are looking for. The Gibbs free energy minimization is an easier path to get and understand the equilibrium concept as in Fig. 1.1. In the same idea, the capability to perform any effect is zero at the equilibrium, with maximum S [kJ/K] or minimal $G = H - TS$ [kJ]. Applying a differential form for Gibbs free energy, with the relation $TdS = dH - \forall dP$, $dG = dH - d(TS) = dH - SdT - TdS = \forall dP - SdT$. Hence, Eq. 1.2 becomes a relation of properties with the objective of minimization.

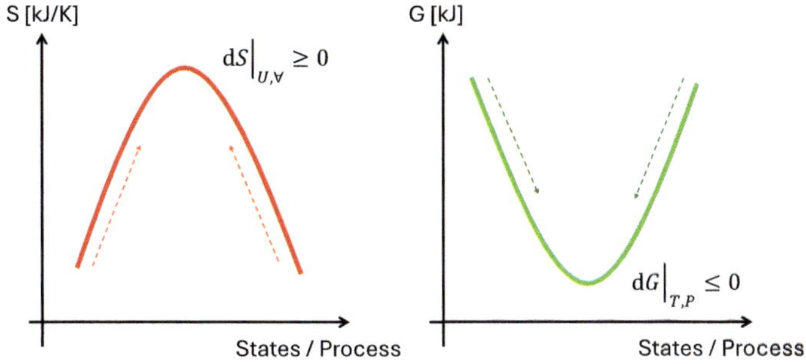

Fig. 1.1 Thermodynamics equilibrium for entropy and Gibbs free energy perspective

$$dG(T, P) = \left.\frac{\partial G(T, P)}{\partial T}\right|_P dT + \left.\frac{\partial G(T, P)}{\partial P}\right|_T dP = \forall dP - SdT \tag{1.2}$$

Note that the concept of equilibrium comes only after the concept of the second law of thermodynamics. In the same path, exergy is at a minimum when there is thermal equilibrium with the environment; therefore, there is no more capability to perform any work ($B = H - H_0 - T_0(S - S_0)$), being a more tangible concept than entropy.

In conclusion, when we have a thermodynamic disequilibrium, Fig. 1.1, there is always some capability to perform some work; therefore, there is an exergy. If we do not use this exergy, every work capability will be destroyed, as shown in the Gouy–Stodola theorem. Therefore, every process tends to lead to an equilibrium, and the seek of using this thermodynamic disequilibrium may be considered the core of the exergy analysis of the energy conversion process.

The lectures of Erwin Schrödinger [7] show a first attempt to understand life from the thermodynamics perspective, where the biological systems are considered a non-equilibrium state with the environment, therefore with lower property entropy than the environment. The author claims that there is a necessity for the body to release entropy to the environment to maintain life itself. In this text, we will show a physical concept of entropy variation of the degradation of nutrients, which is the human metabolism on an entropy basis, interpreted as a form to keep the body out of equilibrium with the surroundings. Later Prigogine and Stengers [8] demonstrated the second law of thermodynamics for open systems and related the irreversibilities and the perception of time, coining the term "arrow of time."

The concept of taxing energy based on its quality, specifically the loss of capacity to perform work, has been discussed by various authors. This relationship is discussed in [9], in which it is emphasized that it would deal with incentives for producers and consumers to utilize energy more efficiently. Due to its rational (Cartesian) framework, this method would not require interference from a free-market system. Finally, as demonstrated by [4] based on exergy analysis, it is possible to propose indicators that quantify the extent to which a process relies on renewable sources (renewability indicator) and exergoenvironmental indicators that assess the quality of energy utilization. In a recent study by [10], the exergy approach was employed to assess resource allocation beyond the scope of monetary systems. Bearing in mind the broader context of solving social problems and promoting social development, exergy analysis proves to be a helpful tool. Moreover, it provides insights into the irreversibility of resource utilization within a specific micro-biome.

Concerning different applications inside energy engineering (bio-thermodynamics and bioheat transfer), one of the few areas whose magnitudes are not strictly deterministic is the study of thermal comfort conditions. This area shows that the relationship between humans and their surrounding environment is influenced by individual anthropometric factors. Since the articles of [11–14] who sought to establish connections between the thermal behavior of the human body and environmental conditions through groundbreaking experiments that were ahead of their time, this area gained attention. As described in [15], traditional literature indicators, such as the Predicted Mean Vote (PMV) and Predicted Percentage of Dissatisfied (PPD) proposed by Povl Ole Fanger, are well-established and included in technical standards such as [16] and [17]. Nevertheless, recent applications considering expanding those indicators using the second law of thermodynamics to relate exergy transferred to the environment and exergy destroyed with such parameters have gained attention in the literature as a supplementary indicator.

Still, sports performance stands out in terms of the relationship between thermodynamics tools and the human body. Exergy analysis and efficiency are being utilized to establish a correlation between sports indicators, such as maximum oxygen consumption and lactate threshold obtainable from the ergospirometric test with the exergy indicators. A direct quote from [18] shows the relevance of this field:

> [...] one field may be very interesting for young researchers, namely the analysis of exergy losses in living organisms. For example, the exergetic efficiency of the vegetation of plants is very low. The reasons behind such an effect are worth investigating. In the human body, too, the exergy losses are considerable. In a mature organism, the total exergy of food is transformed into the considerably smaller exergy of waste materials or heat rejected into the environment. The reduction of these losses might be interesting from the point of view of food consumption and human health.

One aspect to consider is the assessment of society and the quality of energy conversion in its various processes, as highlighted by [19] (from the cradle to the grave). According to the works of [20] and [21] the analysis can be applied to specific sectors, such as transportation, examining the trade-offs between performance and efficiency. The residential sector, as seen in [22], along with its appliances' efficiencies [23]. Eventually, [24] explored the use of solar energy in power plants and its application in food production. Lima and Mady [25] investigated the energy consumption of the gelatin drying process, while [26] analyzed the exergy consumption of caustic soda steam production.

Figure 1.2 shows the representation and scope of the study of exergy analysis of biological systems, from cells to society. This figure shows the different sectors in order to understand how a single being (or a population) needs a massive structure to have its actual consumption standards. This text will try to make this bridge in the last few chapters.

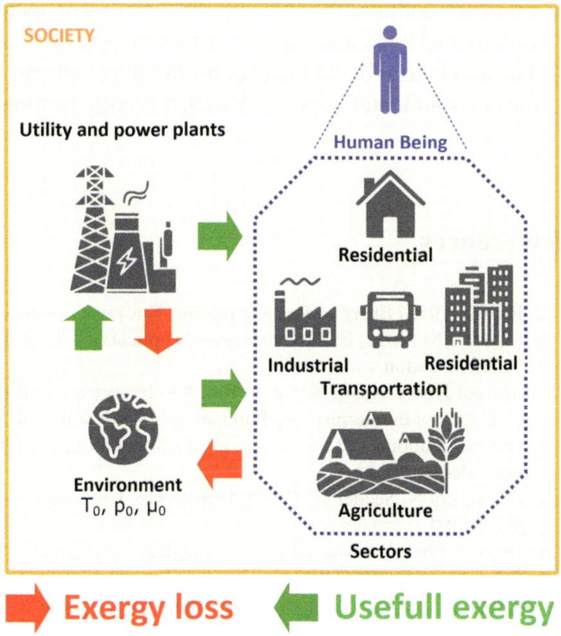

Fig. 1.2 Exergy evaluation under different perspectives

1.2 Text Organization

- Chapter 2 contains an introductory text regarding the second law of thermodynamics, a reference environment, and three applications of this tool in daily routine.
- Chapter 3 consists of the demonstration of the principles of Thermodynamics applied to biological systems, from cellular metabolism to the whole body.
- Chapter 4 describes the adaptation of the human thermal model for the proposition of the exergy efficiency and destroyed exergy as an indicator of sports performance indicators.
- Chapter 5 shows indicators of thermal comfort associated with exergy destroyed and exergy transfer to the environment, differences between anatomies (lean and fat mass), and birth sex (male or female) and the consequences on the thermal response to a more inclusive environment with lower energy consumption.
- Chapter 6, demonstrates the effects of carbon monoxide poisoning on the human respiratory system, and details the exposition of a pregnant women to unhealthy environments such as traffic jam in a tunnel.
- Chapter 7 discusses the effects of cardiovascular pathologies on human energy and exergy behavior.
- Chapter 8 shows the analysis and proposition of different protocols and methods to induce hypothermia in patients after a cardiac arrest, in order to preserve the neurological system of the patient.

- Chapter 9 contains the analysis of the quality of energy conversion in São Paulo, as well as a specific analysis of consumption in the transport and residential sectors. The aim is to provide insights for the development of public policies to decrease the environmental impacts of societies daily routine.

References

1. Rant Z (1955) Energy value and pricing. Strojniski Vestnik 1:4–8
2. Shukuya M (2012) Exergy: theory and applications in the built environment, 1st edn. Springer Nature, London
3. Szargut J, Morris DR, Steward FR (1988) Exergy analysis of thermal, chemical, and metallurgical processes. Hemisphere Publishing Corporation, New York
4. Oliveira-Junior S (2021) Exergy: production, cost and renewability. Springer Science & Business Media
5. De Nevers N, Seader JD (1980) Lost work: A measure of thermodynamic efficiency. Energy 5:757–769
6. Bejan A (2016) Advanced engineering thermodynamics. John Wiley & Sons
7. Schrödinger E (1944) What is life?: the physical aspect of the living cell; with, mind and matter; autobiographical sketches. Cambridge University Press, Cambridge
8. Prigogine I, Stengers I (1997) The end of certainty. Simon and Schuster
9. Hirs G (1993) Exergy loss: a basis for energy taxing. In: Energy efficiency in process technology . Dordrecht: Springer Netherlands
10. Meng Z, Yang Z, Zhang B, Wu X (2023) Resource exergy analysis of the Chinese society in 2017. Environ Sci Poll Res 30(14):42016–42033
11. Harris JA, Benedict FG (1918) A biometric study of human basal metabolism. Proceed Nat Acad Sci 4(12):370–373
12. Hardy JD (1934) The radiation of heat from the human body: III. The human skin as a black-body radiator. J Clin Invest 13(4), 615–620
13. Hardy JD, Du Bois EF, Soderstrom GF (1938) Basal metabolism, radiation, convection and vaporization at temperatures of 22 to 35 C.: Six figures. J Nutrition 15(5):477–497
14. Hardy JD, Du Bois EF (1940) Differences between men and women in their response to heat and cold. Proceed Nat Acad Sci 26(6):389–398
15. Van Hoof J (2008) Forty years of Fanger's model of thermal comfort: comfort for all? Indoor Air 18(3)
16. American Society of Heating, Refrigerating, Air-Conditioning Engineers (2005) ASHRAE handbook
17. ISO7730 (2005) Ergonomics of the thermal environment-Analytical determination and interpretation of thermal comfort using calculation of the PMV and PPD indices and local thermal comfort criteria
18. Szargut J (2005) Exergy analysis. ACADEMIA. Mag Polish Acad Sci 31–33
19. Mosquim RF, Oliveira-Junior S, Mady CEK (2018) Modelling the exergy behavior of São Paulo State in Brazil. J Clean Prod 197:643–655
20. Mosquim RF, Mady CEK (2021) Design, performance trends, and exergy efficiency of the Brazilian passenger vehicle fleet: 1970–2020. J Clean Prod 290:125788
21. Mosquim RF, Mady CEK (2022) Performance and efficiency trade-offs in Brazilian passenger vehicle fleet. Energies 15(15):5416
22. Pereira MTRM, Carvalho M, Mady CEK (2022) Adressing energy demand and climate change through the second law of thermodynamics and LCA towards a rational use of energy in Brazilian households. Entropy 24(11):1524

23. Mady CEK, Pinto CR, Pereira MTRM (2020) Application of the second law of thermodynamics in brazilian residential appliances towards a rational use of energy. Entropy 22(6):616
24. Pinto CR, Mady CEK (2020) Comparing the thermodynamic performance of organic Rankine and Kalina cycles in solar energy systems. J Brazilian Soc Mech Sci Eng 42(11):597
25. Lima LS, Mady CEK (2021) Energy and exergy analysis of an absorption and mechanical system for a dehumidification unit in a gelatin factory. Entropy 23(4):415
26. Mady CEK, Bresciani AE, Paiva JL, Jardin-Junior RN, Guardani R, Oliveira-Junior S (2018) Thermodynamic assessment of MVR implementation in multistage evaporation plants. J Brazilian Soc Mech Sci Eng 40(8):395

Chapter 2
Fundamentals of Second Law and Exergy Analysis

2.1 First and Second Laws of Thermodynamics

The first law of thermodynamics (energy conservation) for any control volume, with several inlets and outlets, considering constant properties in the flow cross-section, can be written as indicated in Eq. 2.1, with the usual hypothesis applied to get to this format from integral equations, and disregarding nuclear, magnetic, electrical, or superficial tension effects. A direct reading of this equation concludes that one can only change the energy of a control volume based on heat transfer rates, execution of a given power, and mass inputs and outputs that carry a given enthalpy (internal energy and flow work), kinetic energy, and potential energy.

$$\frac{d\mathbf{E}_{CV}}{dt} = \sum \dot{m}_{in}\left(h + \frac{V^2}{2} + gz\right)_{in} - \sum \dot{m}_{out}\left(h + \frac{V^2}{2} + gz\right)_{out} + \dot{Q}_{CV} - \dot{W}_{VC} \tag{2.1}$$

In which, $d\mathbf{E}_{VC}/dt$ is the temporal variation of the energy content in the control volume ($\mathbf{E}_{VC}$), $\dot{Q}_{VC}$ is the term referring to the heat transfer rates by the boundary of the control volume, $\dot{W}_{VC}$, the rate of doing work (power) and $\sum \dot{m}\left(h + \frac{V^2}{2} + gz\right)$, the enthalpy, kinetic, and potential energy flow rates that cross the boundary of the control volume, where, in, can refer to the input or out, the output.

The second law assumes the format displayed by Eq. 2.2 for a control volume. Since it is not a conservative law (only in imaginary reversible processes), there is a production or generation of irreversibilities $\dot{\sigma}$ (kW/K). It is only possible to change the entropy of the control volumes S_{CV} by a mass flow rate passing through the surface transporting "a given amount" of the property entropy, s (kJ/kgK), into or out of its boundaries and through heat transfer rates pathways ($\dot{Q}_i/T_i$). It is important to observe that in Eq. 2.2, the heat transfer rate is in the numerator and the temperature in the denominator; hence, a measure of energy, e.g., kJ divided by temperature in Kelvin. The choice of this temperature depends on the person applying the equation;

C. E. Keutenedjian Mady, *Exergy Analysis*, Green Energy and Technology,
https://doi.org/10.1007/978-3-032-01646-1_2

hence, it is the irreversibility ($\dot{\sigma}$), internal or external, of the control surface. We then discuss the useful effect of this rate of energy transfer, i.e., entropy transfer. This last term shows the distance the entropy associated with the heat transfer is from the reversible one if it is a closed system $\Delta S \leq \sum \dot{Q}_i/T_i$, or $\Delta S = \sum \dot{Q}_i/T_i + \sigma$. This result has a profound meaning: for the same entropy variation, the higher $\dot{Q}_i/T_i$, the lower the quality of the heat transfer rate, and the process is more apart from the reversible one.

$$\frac{dS_{CV}}{dt} = \sum \dot{m}_{in} s_{in} - \sum \dot{m}_{out} s_{out} + \sum_i \frac{\dot{Q}_i}{T_i} + \dot{\sigma} \tag{2.2}$$

Prigogine and Stengers [1] discuss that the 19th century gave us two heritages. The first one is related to Newton's law with no friction, which is a static universe, although it brings a description related to the word entropy. Entropy is an essential property introduced by thermodynamics, bringing light to irreversible processes and, therefore, being time-oriented (always positive). The compatibility with time brought the nickname "arrow of time."

In a period in which there was a need for a concept to give the maximum available work for a given process, Clausius enunciated the "second principle of thermodynamics". The first law was limited to verification of cycle behavior, or energy conservation principle, $\oint \delta \dot{Q} = \oint \delta \dot{W}$, hence $\dot{W}_{liq} = \dot{Q}_H - \dot{Q}_L$, where $\dot{Q}_i$ is the heat transfer rate from T_i and $\dot{W}$ is the power output. Clausius shows the world an equation that is non-conservative by nature, for a thermodynamic cycle: $\oint \delta \dot{Q}/T \leq 0$, hence, $\dot{Q}_H/T_H - \dot{Q}_L/T_L \leq 0$. If we modify the inequality of Clausius to a process from state 1 to 2, it becomes $\Delta S \geq \int_1^2 \delta Q/T$, which is the second law principle.

Likewise, the statement: "It is impossible to construct a device that operates in a thermodynamic cycle and produces no effect other than the transfer of heat from a lower-temperature thermal reservoir to a higher-temperature thermal reservoir." Therefore, Fig. 2.1 shows that there must be some work as input, and the performance coefficient cannot tend to inf.

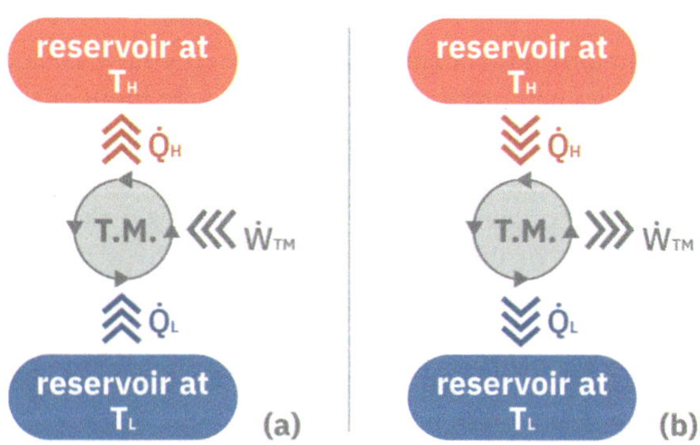

Fig. 2.1 Thermal machines operating in thermodynamic cycle for **a** refrigeration cycle and **b** power cycle, between a reservoir at T_H, or higher temperature than the reservoir at T_L, or lower temperature

For a refrigeration cycle (Fig. 2.1a), the performance coefficient is calculated by Eq. 2.3.

$$COP = \frac{\dot{Q}_L}{\dot{W}_{liq}} \tag{2.3}$$

Further, the Kelvin-Planck statement goes similarly: "no device can operate in a thermodynamic cycle communicating with only one thermal reservoir and produce solely work". In other words, there must be contact through heat transfer with a reservoir with lower temperatures, meaning that the energy efficiency (Eq. 2.4) will never be 100%, always $\dot{W}_{liq} < \dot{Q}_H$.

$$\eta_{cycle} = \frac{\dot{W}_{liq}}{\dot{Q}_H} \tag{2.4}$$

Hence, for an irreversible process as in [2–4], the line integral of the ratio heat (δQ) divided to the temperature (T), over a cycle may never be higher than zero (Eq. 2.5). For a reversible process, it is equivalent to zero but never higher. This statement is profound in its structure that can be interpreted as, although the energy in its several forms of transfer is conservative for any given process, the quality of the energy never is! It only decreases to a quality equal to heat transfer to the lower thermal reservoir, Q_L/T_L. This non-conservative nature may bring light to the definition of Shannon entropy, and the possible macro-states, which are sometimes confused only as a "degree of disorder of the system" or, more adherent to Shannon's proposition, the probability for the energy to spread.

$$\oint \frac{\delta Q}{T} \leq 0 \tag{2.5}$$

If we open these equations in a reversible and irreversible process, in a closed system between 1 and 2, the property entropy appears by itself, which is $\Delta S = \int_1^2 \frac{\delta Q}{T}|_{rev}$. Therefore, the property entropy (s, [kJ/kgK]) is the reversible heat transfer, the one with the necessity of an infinitesimal temperature gradient to happen (δT). Therefore, everything other than this reversible heat is called irreversibilities (or entropy generated, by some textbooks) which is not a property but a function of the process ($\Delta S \leq \int_1^2 \frac{\delta Q}{T}$). As indicated in Eq. 2.6 for a closed system, the irreversibilities, ${}_1\sigma_2$ [kJ/K], depend on the path of the systems for fixed states 1 and 2.

$$S_2 - S_1 = \int_1^2 \frac{\delta Q}{T} +{}_1\sigma_2 \tag{2.6}$$

To generate or destroy something is not usual for natural laws, which makes the core of the second law of thermodynamics. Here is where the theorem of Gouy–Stodola makes sense for engineers. Although because of its unity [kJ/K], the irreversibilities may have a problematic meaning, it shows how much of the quality of the energy is lost and how far it is from the ideal-reversible energy conversion process. The entropy generations show two different scenarios of energy process conversion, which leave more trace elements in the environment, whereas the Lost Work shows that because of the irreversibilities, some amount of work is destroyed, $W_{lost} = T_0\sigma$ [kJ], in an energy conversion process. In this section, the references used were [5–7].

Wherever it is necessary to use a combination of the first and second laws of thermodynamics, using Gibbs expressions is a proper tool. The deduction is simple, which is the application for the first law of thermodynamics for a closed system ($dU = \delta Q - \delta W$); when the work is reversible $\delta W = Pd\forall$, and for a reversible heat transfer $\delta Q = TdS$, this relation always functions to compare thermodynamic states, being only a function of point functions. Hence, it is a general expression of state or state variation between two thermodynamic states independently of the process. Equations 2.7 and 2.8 show these properties' relations.

$$TdS = dU + Pd\forall \tag{2.7}$$

$$TdS = dH - \forall dP \tag{2.8}$$

There are some simplifications, that for a incompressible solid or liquid $du \approx dh \approx cdT$, there is no more distinction between c_p and c_v. Hence, $S_2 - S_1 = cln\,(T_2/T_1)$. Another particular case is for ideal gases, where $du = c_v dT$ and $dh = c_p dT$, where the equation of state may be written as a function of the specific volume $v = \forall/m$ in [m^3/kg], hence $pv = R_g T$, it is possible to obtain Eqs. 2.9 and 2.10.

$$S_2 - S_1 = c_v ln\left(\frac{T_2}{T_1}\right) + R_g ln\left(\frac{\forall_2}{\forall_1}\right) \tag{2.9}$$

$$S_2 - S_1 = c_p ln\left(\frac{T_2}{T_1}\right) - R_g ln\left(\frac{P_2}{P_1}\right) \tag{2.10}$$

2.2 Exergy Analysis

The exergy analysis consists of the simultaneous use of the first and second Laws of thermodynamics (linear combination) in the evaluation of the quality of an energy conversion process, which, according to Oliveira-Junior [7], allows an effective thermodynamic evaluation of processes of energy conversion, since it quantifies the irreversibilities that occur during their development, in another thermodynamic basis, the destruction of the capability to perform work. Szargut et al. [3] defines exergy as the maximum work that can be performed when a system evolves from the state it is in to the state of equilibrium with the components of the environment (mechanical, thermal, and chemical), through reversible processes, interacting only with the elements of the environment. The first to use the nomenclature was Rant [8].

2.2.1 Necessity of Reference Environment

The inspiration for this section is related to an exercise from the textbook of [9]. The problem: "We have at our disposal three cylinders of pure helium gas as in Fig. 2.2".

These may be designated as A, B, and C. The initial temperatures, pressures, and volumes are given by Table 2.1. We allow interactions among the cylinders. The c_p of the helium may be considered as 20.9 J/(mol.K). Helium may also be considered an ideal gas.

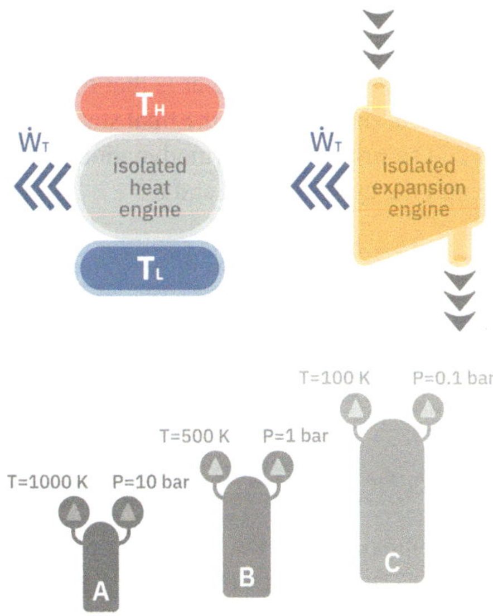

Fig. 2.2 Heat engines that produce work from heat transfer between two cylinders or produce work from a difference of pressure between the two cylinders Obtained and adapted from [9]

Table 2.1 Initial thermodynamic states. Obtained from [9] and calculated final state using the first and second law of thermodynamics

STATE 1 Cylinder	∀ [m^3]	T [K]	P [kPa]	m [kg]
A	0.01	1000	1013	0.004878
B	0.05	500	101.3	0.004878
C	0.1	100	10.13	0.004878
STATE 2 Cylinder	∀ [m^3]	T [K]	P [kPa]	m [kg]
A	0.01	288.5	54.82	0.0009144
B	0.05	288.5	54.82	0.004572
C	0.1	288.5	54.82	0.009144

What is the *maximum* amount of work that can be produced? Heat engines, expansion engines, etc., are allowed; the only restraint is that there must be no effect on the environment other than the rise of weights in the gravitational field (i.e., work reservoir). This would then imply that the environment would not be compressed or expanded; that is, the total volume of the three cylinders must be equal to their combined volumes at the end.

It is my opinion that there is a purpose to this exercise. The difficulty of not having a final thermodynamic state, or a reference environmental state (T_0, P_0). In this exercise, it does not matter where the process occurs: earth, mars, moon. This brings light into the second law analysis since all the processes are reversible and have no environmental interaction. Moreover, there may be a disruption in reality since there is no final natural state. Indeed, it is possible to achieve it, though what is the meaning of this final state? Another point is that the mass and volume of the three cylinders are the sums of the three, and the final state is equal for all cylinders, defined by the number 2. The first law of thermodynamics gives us Eq. 2.11, where $\Delta U = (U_2 - U_1)_i$, being i the cylinder A, B and C:

$$(U_2 - U_1)_A + (U_2 - U_1)_B + (U_2 - U_1)_C = Q_{system} - W_{system} \tag{2.11}$$

Considering $Q_{system} = 0$, and for perfect gases, the specific heat at constant volume is $c_v \equiv du/dT$, and it does not change with relation to temperature ($R_g = c_p - c_v$), Eq. 2.11, becomes Eq. 2.12.

$$m_{A,1}c_v(T_2 - T_1)_A + m_{B,1}c_v(T_2 - T_1)_B + m_{C,1}c_v(T_2 - T_1)_C = -W_{net} \tag{2.12}$$

The performed work is maximum if the process is reversible, and in this case, adiabatic. Therefore, we must find the final temperature and pressure of the process.

The second law for adiabatic and reversible processes gives, Eq. 2.13. The original form is in Eq. 2.6, with no heat transfer to the environment ($Q_{system} = 0$) and no irreversibilities (${}_1\sigma_2$=0).

$$(S_2 - S_1)_A + (S_2 - S_1)_B + (S_2 - S_1)_C = 0 \tag{2.13}$$

Since there is no heat transfer, neither irreversibilities ($W_{MAX} = W_{net}$). Using Gibbs function $TdS = dH - \forall dP$, for a gas with no change in the mass and volume, it is possible to obtain $\Delta S = S_2 - S_1 = c_p ln(T_2/T_1) - R_g ln(P_2/P_1)$, with $P\forall = nR_gT$ (Clapeyron equation). The final state is a system of equations (second law and state equation), where $T_2 = 288.5$ K, $P_2 = 54.82$ kPa and $W_{MAX} = 11.27$ kJ.

What if these cylinders were on the surface of the moon, mars, or earth? It would not matter. Since all thermodynamic problems are in the environment, and the heat engines must have contact with the environment, the same for the expansion engines. The exergy comes with this sense of reality, i.e., a reference final state we live in. The following sections are written inspired by this problem and how the exergy analysis contributes to a better understanding of the maximum available work, with an actual reference environmental (dead state). In both cases, it is essential to highlight that the calculated maximum available work is a property since it only depends on two thermodynamic states and the environment (for pure substances). Adopting the dead-state to be $T_0 = 298$K and $P_0 = 101.325$ kPa, and the heat turbine exchanges heat with the environment at T_0, we have the $W_{MAX} = 10.82$ kJ. Which is lower than the one produced by the exercise; nevertheless, it is the actual maximum available work taking into account a reference environment. A deeper discussion of this idea was carried out by Gallo and Milanez [10].

2.2.2 Exergy Analysis: Physical and Chemical Exergy

Figure 2.3 represents a simplified control volume inside the environment, with a finite difference of temperature ($T_{CV} > T_0$), pressure ($P_{CV} > P_0$), and chemical potential ($\mu_{CV} > \mu_0$). Therefore, the exergy content inside the control volume ($\mathbf{B}_{CV}$) is distinct from the exergy of the environment ($\mathbf{B}_0 = 0$). In this specific control volume, only one flow rate enters and leaves the control surface since it is an open system.

In Fig. 2.3 there is a representation of a control volume (CV), there are also two energy transfer rate forms: (i) heat transfer rate (from the control volume at T_{VC} to the environment at T_0) and (ii) performed power (pure exergy). The exergy flow rate is obtainable from the combination of the first and the second laws of thermodynamics $\dot{m}_i b_i = \dot{m}_i[h_i - h_0 - T_0(s_i - s_0)]$, where $i = in, out$. Equation 2.14 shows the exergy analysis applied to the control volume

$$\frac{d\mathbf{B}_{CV}}{dt} = \dot{m}_{in} b_{in} - \dot{m}_{out} b_{out} + \dot{Q}_{CV}\left(1 - \frac{T_0}{T_{CV}}\right) - \dot{W}_{CV} - \dot{B}_{dest} \tag{2.14}$$

In order to assess Eq. 2.14, the application of the first law of thermodynamics, for Fig. 2.3, leads to Eq. 2.15 considering steady state, no significant variation of kinetic and potential energies from the inlet to the outlet of the control volume.

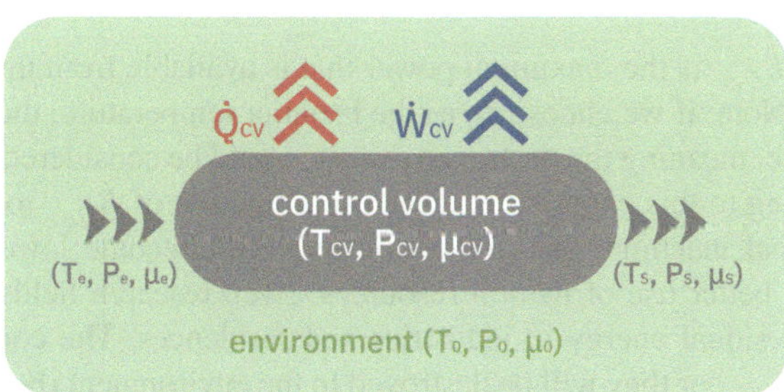

Fig. 2.3 A simplified control volume inside an environment. The subscripts *in*, *out* refer to input and output. The terms H, S, and μ are, respectively, enthalpy, entropy, and chemical potential. *Source* based and adapted from [2]

$$0 = \dot{m}\,(h_{in} - h_{out}) + \dot{Q}_{CV} - \dot{W}_{CV} \tag{2.15}$$

To apply the second law of thermodynamics, one must state the frontier of the systems. Is it at T_0 or T_{CV}? From the second law standpoint, it isn't easy to explain that; nevertheless, if one chooses the prime, all the irreversibilities associated with the heat transfer with a finite temperature difference will be internal to the system (an imaginary frontier at T_0), if one considers the latter part of the irreversibilities considered as being generated outside of the open system, which will be named external irreversibilities. Equation 2.16 shows the result using T_{CV} as the frontier, being this temperature the average of the control surface.

$$0 = \dot{m}\,(s_{in} - s_{out}) + \frac{\dot{Q}_{CV}}{T_{CV}} + \dot{\sigma}_{CV} \tag{2.16}$$

Multiplying the second law (Eq. 2.16) by $\times(-T_0)$ and combining it with the first law (Eq. 2.15) the result is Eq. 2.17, which is the same as Eq. 2.14, for steady-state conditions, one mass flow entering and leaving the system.

$$0 = \dot{m}\,[h_{in} - h_{out} - T_0\,(s_{in} - s_{out})] + \dot{Q}_{VC}\left(1 - \frac{T_0}{T_{CV}}\right) - \dot{W}_{CV} - \dot{\sigma}_{CV} \times T_0 \tag{2.17}$$

Changing $\dot{\sigma}_{CV} \times T_0$ to $\dot{B}_{dest}$, it is possible to obtain Eq. 2.18.

$$\dot{B}_{dest} = \dot{m}\,[h_{in} - h_{out} - T_0\,(s_{in} - s_{out})] + \dot{Q}_{VC}\left(1 - \frac{T_0}{T_{CV}}\right) - \dot{W}_{CV} \tag{2.18}$$

Two aspects already cited over the text arise from Eqs. 2.17 and 2.18. First, Frontier's choice changes the irreversibilities and $\dot{B}_d = \dot{\sigma}_{CV} \times T_0$, which is the Gouy–Stodola theorem known as **lost work** now named destroyed exergy rate. Another matter is the natural result is the Carnot maximum work, Eq. 2.19

$$\dot{B}_{\dot{Q}_{CV}} = \dot{W}_{MAX} = \dot{Q}_{CV}\left(1 - \frac{T_0}{T_{CV}}\right) \tag{2.19}$$

In Eq. 2.19, $\dot{B}_{\dot{Q}_{CV}}$ is the maximum power that is available from this heat transfer at T_{CV} and T_0. Note: if we choose T_0 as the Frontier temperature, the Carnot factor will become zero, meaning the destroyed exergy would be considered internal to the system. Returning to this expression, if we use a fraction of $\dot{B}_{\dot{Q}_{CV}}$ to perform work through a thermal machine, this irreversibility would become lower, with higher efficiencies and better use of natural resources. Even research fields are emerging from utilizing residual energy in industries and residences. The concept is to use these exergies $\dot{B}_{\dot{Q}_{CV}}$, or they will be destroyed in the environment (boundary at T_{CV}) or inside the control volume (boundary at T_0).

2.2.2.1 Physical Interpretation of the Exergy Associated with a Heat Transfer Rate

Herein we will discuss for the first time the necessity of having a fixed environment, or as in the textbooks: a thermal reservoir [2]. Considering Eq. 2.20, what would be the maximum power attainable from a heat transfer rate ($\dot{Q}_{CV}$) between a temperature higher than the environment (T_0) to the environment? And if the temperature is lower than the environment. What is the meaning of this term or the expected result?

$$\dot{B}_{\dot{Q}_{CV}} = \dot{Q}_{CV}\left(1 - \frac{T_0}{T_{CV}}\right) \tag{2.20}$$

Figure 2.4 indicates the interpretation of the exergy associated with a heat transfer for the case that there is a thermal reservoir (thermal capacity tending to infinity) with a temperature higher than the environment. This finite temperature difference ($\Delta T = T_{CV} - T_0$) between these two reservoirs shows some availability. Hence, Carnot's ideal thermal machine converts the heat transfer rate into the maximum achievable power with Carnot Efficiency (or Carnot Factor). Now, if we let nature take its course, this availability of work is lost [11], and this exergy $\dot{B}_{\dot{Q}_{CV}}$ is destroyed.

The exergy of a heat transfer is used to compare and measure efficiency concerning the Carnot factor, for instance. If we use any other thermal cycle, it is possible to obtain Eq. 2.21, assuming $\dot{W}_{CV}$ as the power of a real process with lower performed power than Carnot's. In this Equation $\dot{W}_{CV}$ is the real work of any cycle, η_{en} is the energy efficiency of the cycle and η_b is the exergy efficiency of the cycle, η_{Carnot} is the Carnot efficiency.

$$\eta_b = \frac{\dot{W}_{CV}}{\dot{B}_{\dot{Q}_{CV}}} = \frac{\dot{W}_{CV}}{\dot{Q}_{CV}}\left(1 - \frac{T_0}{T_{CV}}\right) = \eta_{en} \times \eta_{Carnot} \tag{2.21}$$

A different situation happens when there is a finite difference in temperature ΔT, and one reservoir has a temperature lower than the environment, e.g., inside a freezer

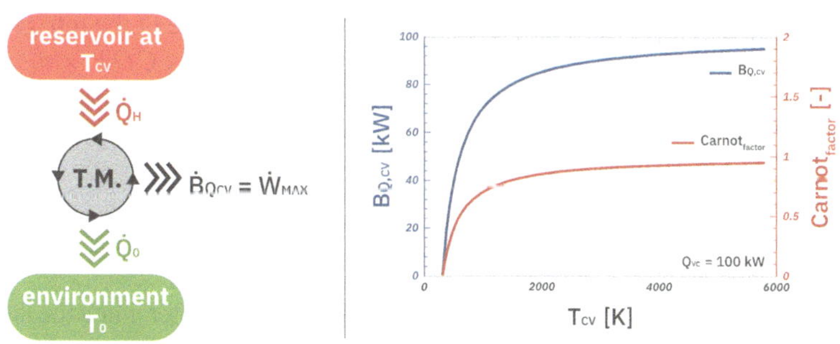

Fig. 2.4 Exergy associated with a heat transfer rate from $T_{CV} > T_0$. Result os exergy and Carnot Factor (limited to 1). *Source* Own authorship based on [2]

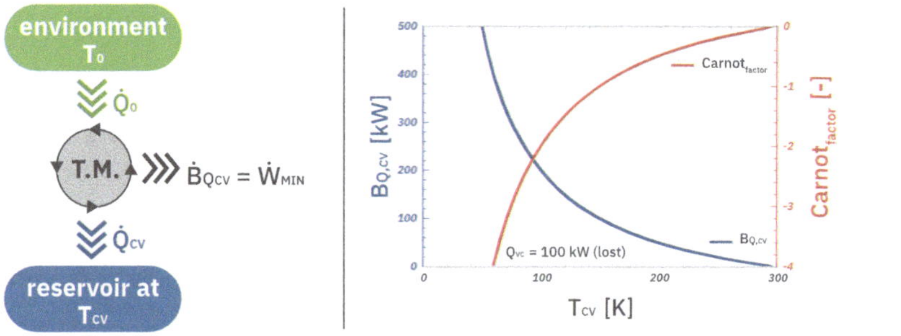

Fig. 2.5 Exergy associated with a heat transfer rate from $T_{CV} < T_0$. Results of exergy and Carnot factor (not limited to 1). *Source* Own authorship based on [2]

with constant temperature. The natural outcome, similar to Fig. 2.4, is that heat would be transferred from T_0 to T_{CV}; nevertheless, to maintain the reservoir at a constant temperature, we can imagine a Carnot refrigeration cycle that removes the heat from the reservoir at T_{CV}, being Q_{CV} negative. At the same time, in Fig. 2.5, Carnot's efficiency becomes negative and achieves values higher than unity in magnitude, showing the amount of exergy necessary to obtain the refrigeration effect. Nevertheless, the exergy associated with heat is always positive. Why? This relation shows that the environment with a lower temperature than the actual environment has higher exergy. Mathematically, it is a product of two negative thermodynamic quantities. When the evaporator removes heat from the "cold" environment, it increases the exergy of the environment. The signs of heat and exergy are different. Hence, the air's exergy content in the freezer is higher than in the environment. If we aim to keep it with lower temperatures than T_0, removing heat from the first law point of view is necessary, but how? It occurs by transferring an exergy rate to a cold area using a thermodynamic cycle, usually using a refrigeration cycle that consumes electrical energy.

The importance of an indicator lower than one may be seen in some air-conditioning announcements using relative efficiency, which is the ratio of the real COP to the Carnot COP. Equation 2.22 shows the exergy efficiency of the conditioning system, where the COP is the coefficient of performance of the refrigeration cycle, which can be higher than one. Additionally, the product $COP \times \eta_{Carnot}$ is never lower than unity, showing a scale of efficiency based on exergy analysis (not possible to obtain solely from the first law of thermodynamics).

$$\eta_b = \frac{\dot{B}_{\dot{Q}_{CV}}}{\dot{W}_{CV}} = \frac{\dot{Q}_{CV}}{\dot{W}_{CV}}\left(1 - \frac{T_0}{T_{CV}}\right) = COP \times \eta_{Carnot} \tag{2.22}$$

Figure 2.6 shows the exergy flow associated with a heat transfer when the temperature is lower than the environmental one, showing that when one removes heat from the reservoir with lower temperatures, there is an exergy provided to maintain the reservoir at lower temperatures than the environment, therefore with a higher exergy.

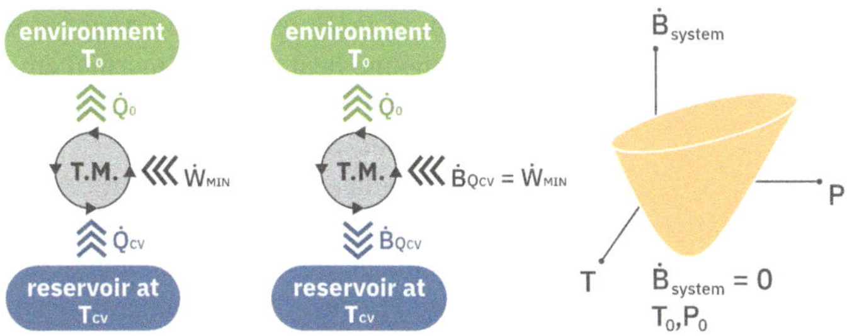

Fig. 2.6 Difference between the thermodynamic cycle with energy flows and exergy flows associated to the refrigeration process

It is necessary to state that the exergy of a system is always $B_{system} \geq 0$. Therefore, the heat is removed from the lower reservoir, and its exergy increases.

2.2.2.2 Physical Interpretation of the Exergy of a Flow Rate

Equation 2.23 shows the flow rate with a given thermodynamic state, bearing in mind that e_k and e_p are the kinetic and potential energy/exergy. Then the stream enters or leaves out of a system at state i, with a pressure (P_i), temperature (T_i), and chemical potential (μ_i). The environment is considered at a pressure (P_0), temperature (T_0), and chemical potential (μ_0).

$$\dot{B}_i = \dot{m}_i b_i = \dot{m}\left[e_k + e_p + h_i - h_0 - T_0\left(s_i - s_0\right)\right] \tag{2.23}$$

Therefore, each flow rate carries a kinetic exergy (same as energy due to its quality) associated with the square of the velocity (a mechanical characteristic of the flow), potential exergy (and energy due to its quality) associated with height (ΔZ) (considering g the gravitational constant). In most engineering applications, these terms are negligible compared to other energy transfer aspects. However, this may not be true for the human body, e.g., jogging (certain V or a ΔV) or climbing stairs (ΔZ). Usually, $e_k = V^2/2$ and $e_p = gZ$ are omitted in this chapter. Nevertheless, they appear in several applications. Bearing in mind that nuclear, magnetic, electrical, and surface tension are not significant.

From a chemical–physical point of view, it has a mechanical and thermal capability to perform some power, hence exergy in comparison to the environment (T_0, P_0, μ_0). There may be several interpretations of this process; nevertheless, [2] tried to systematize it from a mechanical perspective. Moreover, these are thermodynamic properties; the path in which we change the properties in reversible processes, only communicating with a thermal environment, does not affect the result.

1. The first is bringing from T_i to T_0 with thermal machines, based on "micro" Carnot's engines, that use each infinitesimal step lower temperature difference to deliver power until $T_i = T_0$, therefore there is no more thermal exergy.

2. The second is getting the pressure from P_i to P_0 using isothermal turbines ($P_i > P_0$) or compressors ($P_i < P_0$) that exchange heat with the thermal environment. Here is the end of all mechanical and thermal processes; there is no more physical exergy, and the system achieves a **restricted reference state**.
3. The final step is more complicated and is related to reversible processes using the van't Hoff equilibrium box. Herein, there is a remarkable proposition described in [2] to combine the mechanical and chemical interpretations, where all reactions occur in equilibrium conditions at T_0 and P_0, therefore not generating any entropy, although exchanging heat ($\dot{Q}_0$) with the environment (T_0). This method describes how the components of the system achieve their composition equal to the actual environment after the complete degradation of chemical materials that belong to this environment, hence $\mu = \mu_0$ **dead state, equilibrium with environment**.

After these steps, this point is called the dead state (there is no possibility of obtaining any work, since the thermodynamic system is at thermal, mechanical (hence physical) and chemical equilibrium with the environment.

Physical exergy–thermal share

Suppose a thermal process that brings the flow rate from T_i to T_0, holding the pressure constant at P_i; what would be the performed power and the physical interpretation of Eq. 2.23, in a general form? With this equation, it is hard to picture that the temperature and pressure are independent. Nevertheless, we must remember that these are thermodynamic properties, and the path in which some processes occur does not matter for the same initial and final states (mathematically, it is a point function).

$$\dot{B}_{*}^{\Delta T} = \dot{m}_{*} b_{*} = \dot{m}\left[h(T_i, P_i) - h_{*}(T_0, P_i) - T_0\left(s_i(T_i, P_i) - s_{*}(T_0, P_i)\right)\right] \quad (2.24)$$

First, for the sake of simplicity, consider a perfect gas with constant specific heat, c_p [kJ/kg.K]. From Joule's experiment, this is a typical case of $u = u(T)$ and $h = h(T)$; these properties are a function solely of the temperature (hypothesis of perfect gas). The entropy depends on the temperature and pressure $s = s(T, P)$. The flow rate exergy would be in Eq. 2.24. The state represented by the sub-index "$_*$" is considered as an intermediate state where the temperature is T_0, with the same initial pressure P_i and chemical potential μ_i.

From the definition of specific heat at constant pressure for the perfect gas c_p, the enthalpy change becomes $h_i(T_i) - h_0(T_0) = c_p(T_i - T_0)$. The entropy variation comes from Gibbs relations, choosing $Tds = dh - vdp$, by applying the hypothesis of perfect gas: $dh = c_p dT$ and $v = R_g T/P$, hence Eq. 2.25 represents the entropy variation of an ideal gas from the actual state until the thermal and mechanical equilibrium with the environment (physical equilibrium).

$$s_i(T_i, P_i) - s_0(T_0, P_0) = c_p ln\left(\frac{T_i}{T_0}\right) + R_g\left(\frac{P_i}{P_0}\right) \quad (2.25)$$

If the system evolves from T_i to T_0 with constant pressure P_i, and this change of thermodynamic property is in Eq. 2.26.

$$s_i(T_i, P_i) - s_*(T_0, P_i) = c_p\left(ln\frac{T_i}{T_0}\right) \tag{2.26}$$

Combining the first and second law of thermodynamics gives us another perspective, which is the maximum available work possible to obtain from state i to state $*$. Hence Eq. 2.27 gives the thermal exergy for an ideal gas. An expression that can never be lower than zero.

$$\dot{B}^{\Delta T}_{i\to *} = \dot{m}c_p\left[T_i - T_0 - T_0 ln\left(\frac{T_i}{T_0}\right)\right] \tag{2.27}$$

Figure 2.7 indicates the thermal exergy and its physical understanding. For a temperature higher than the environment temperature (T_0) it is possible to imagine several Carnot's engines that use differential energy from the flow ($\delta Q_{H,i} = dH_{fluid} = H(T_i, P_i) - H_*(T_0, P_i)$). Instead of dispatching and destroying this exergy into the environment, it is possible to imagine these thermal machines extracting the maximum power. This type of exergy may cause some confusion with the exergy associated with heat transfer, although it is clear that an enthalpy change results in thermal

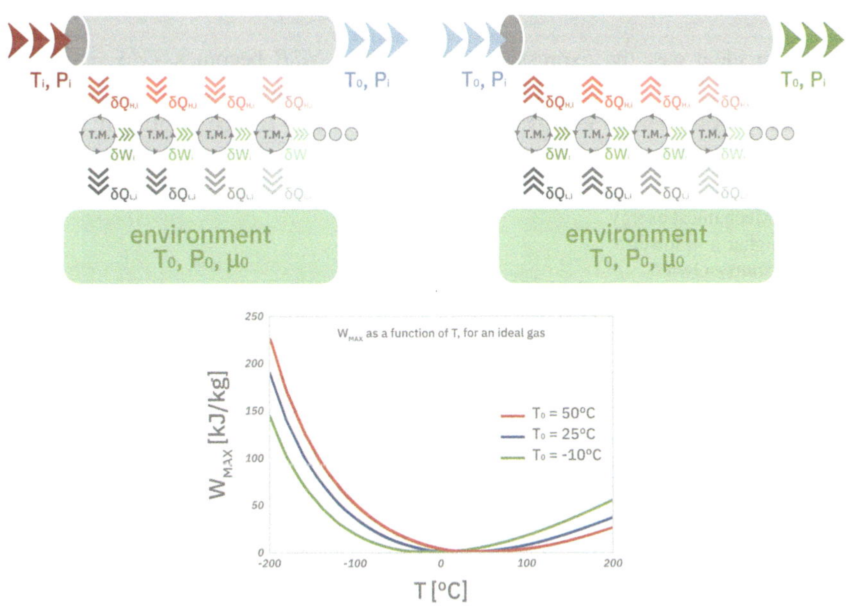

Fig. 2.7 Thermal exergy from T_i higher **a** or lower **b** than T_0. *Source* Own authorship based on [2]

exergy. In contrast, it is necessary to remember that there is a mass flow rate carrying some exergy with temperature and pressure modification, not solely a reservoir at a particular temperature.

Therefore, for any fluid or a specific gas, the idea is to obtain the maximum work ($\delta \dot{W}$) from a heat transfer ($\delta \dot{Q}$). This idea becomes relevant when the discussion is rational use of the energy as in [12] or even using low quality energy like in Organic Rankine Cycles (ORC) [13].

Physical exergy–mechanical share

Figure 2.8 indicates the exergy content in a flow rate with a pressure higher ($P_i > P_0$) or lower ($P_i < P_0$) than the environmental pressure (P_0). Therefore, as in the two machines of Fig. 2.2 (exercise used to understand the necessity of reference environment), the heat and expansion engine are a path to the actual maximum available work; nevertheless, when we use exergy, there must be heat exchanges with the environment, which is at T_0 and P_0. In Fig. 2.8, the expansion engines have to suffer some modifications to a non-adiabatic version of the turbines (heat transfer from the environment) and compressors (heat transfer to the environment).

Equation 2.28 indicates the maximum work that can be obtained from P_i to P_0, or as minimal work as input from taking P_i to P_0 content. A general expression

$$\dot{B}_{i \to *}^{\Delta P} = \dot{m} \left[h_*(T_0, P_i) - h(T_0, P_0) - T_0 \left(s_*(T_0, P_i) - s(T_0, P_0) \right) \right] \tag{2.28}$$

For an ideal gas, the expression $TdS = dH - \forall dP$ becomes $TdS = -\forall dP$, therefore, Eq. 2.28 becomes 2.29 simplified to an isothermal process of an ideal gas. Differently from thermal exergy, the mechanical exergy can be positive if $P_i > P_0$

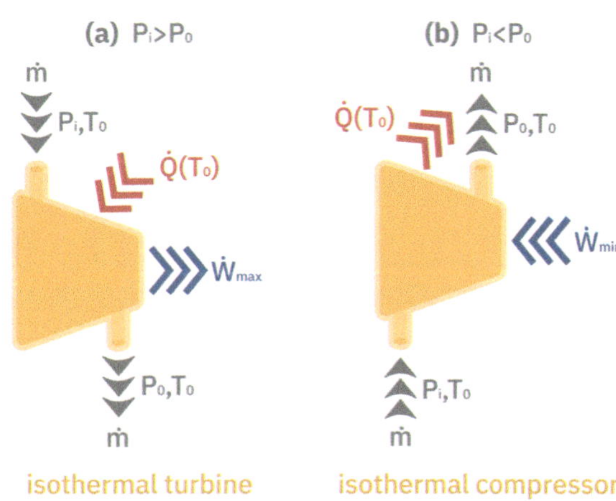

Fig. 2.8 Mechanical exergy from P_i higher **a** or lower **b** than P_0. *Source* Own authorship based on [2]

or negative if $P_i < P_0$ the heat exchanges with the environment are to maintain a constant temperature.

$$\dot{B}^{\Delta P}_{*\to 0} = \dot{m} R_g T_0 ln \left(\frac{P_i}{P_0}\right) \tag{2.29}$$

The total physical exergy is the combination of both terms, as in Eqs. 2.7 and 2.29: $\dot{B}^{ph} = \dot{B}^{\Delta T} + \dot{B}^{\Delta P}$, for an ideal gas is in Eq. 2.30 in [W] or [kW].

$$\dot{B}^{ph} = \dot{m} cp \left[T - T_0 - T_0 ln \left(\frac{T}{T_0}\right)\right] - \dot{m} R_g T_0 ln \left(\frac{P}{P_0}\right). \tag{2.30}$$

Chemical exergy

Recapitulating, the physical exergy is the maximum available amount of work when a system or substance flow passes from an initial state to the restricted reference state, characterized by thermal and mechanical equilibrium with the environment. The chemical exergy is the maximum work that is possible to obtain when a system or flow is taken reversibly interacting with components of the environment from the restricted reference state (T_0, P_0 and μ_0^*) to the dead state (T_0, P_0 and μ_0), characterized by the existence of thermal, mechanical and chemical equilibrium. Therefore, the substance is in equilibrium with the environment.

To calculate the chemical exergy of a substance, one must define a standard environment model, as verified by [2, 3, 7]. In this environment, the following reference substances are considered for a state of thermomechanical equilibrium with T_0 = 298.15 K and P_0 = 101.325 kPa (restricted reference state):

- gaseous components of the atmosphere (O_2, N_2, CO_2, H_2O, among others);
- reference solid substances in the earth's crust, as conducted by [14];
- ionic substances from sea references;
- reference substances in molecular form, non-ionizable from the seas.

As discussed by Valero and Valero [14], now there is a difference in considering the earth as the dead state and the condition that the earth would have without disequilibrium with the universe. One of the most used models for the reference environment is from Szargut et al. [3] and modifications that occurred over time [10, 14, 15]. According to [16], if a reference state of the Earth is defined, the chemical exergy share of an ore, for example, will be equal to zero (if the composition of the ore is equal to the average composition of the Earth) or equal to a value computed based on the composition of the ore, its physical state and Gibbs energy of formation of its constituents. Figure 2.9 illustrates the concept of physical and chemical exergy for any system (or mass flow), in which the physical exergy is the maximum available work to the system going from T and P to T_0 and P_0. At this point, it is like eating a bar of chocolate in environmental conditions. There is an opportunity to obtain the exergy content of the chemical bonds. The maximum work available is the chemical

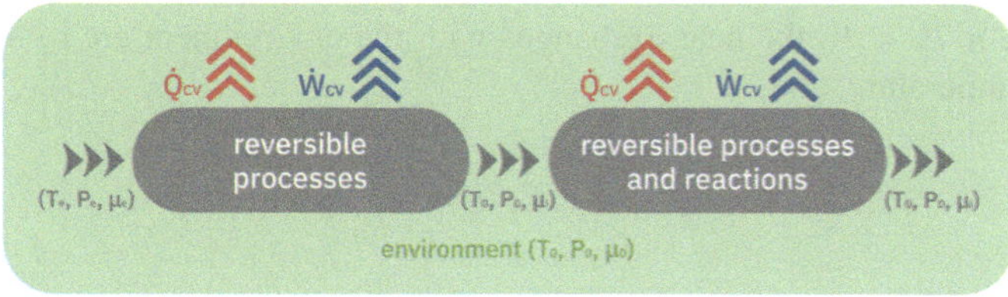

Fig. 2.9 Representation of physical and chemical exergy for a given thermodynamic system

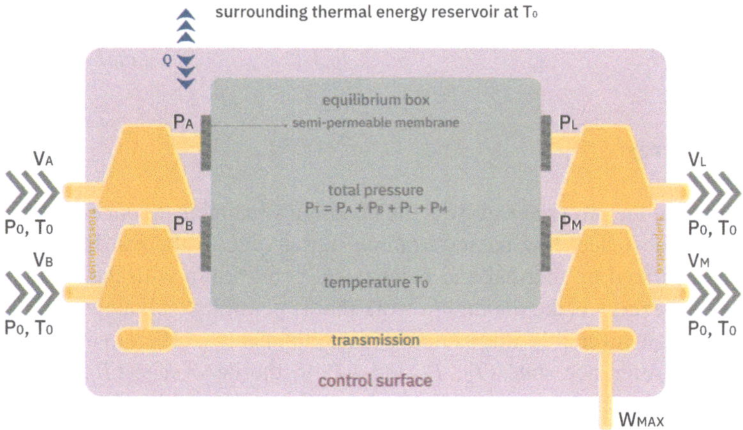

Fig. 2.10 Van't Hoff equilibrium box. *Source* Own authorship based on classical images from [2]

exergy when bringing each component to the respective environmental reference compositions (μ^* to μ_0).

Figure 2.10 shows an idealization of a reaction that occurs in equilibrium, in which substances enter a T_0 and P_0. First, the reactants suffer compression to their partial pressure at the equilibrium box, and the substances formed suffer expansion from their partial pressure to the temperature and pressure of the environment (T_0 and P_0). This reaction is fundamental because of its reversibility and must exchange heat with the environment at $\dot{Q}_0$. Equation 2.31 indicates the reaction that occurs in the box, where there is a free energy variation of $\Delta\bar{g}(T_0, P_0) = \nu_L\bar{g}_L(T_0, P_0) + \nu_M\bar{g}_M(T_0, P_0) - \nu_A\bar{g}_A(T_0, P_0) - \nu_B\bar{g}_B(T_0, P_0)$, which is the maximum available work of a reaction, where all substances enter a T_0 and P_0 and leave at the same thermodynamic state. The bar above the physical quantity, e.g., $\bar{g}$ denotes a molar property [kJ/kmol].

$$\nu_A A + \nu_B B \leftrightarrow \nu_L L + \nu_M M \tag{2.31}$$

Hence, each compressor/turbine that takes a substance from P_0 and T_0 and leaves the reactants to the reaction chamber (P_A, P_B), in the case of Eq. 2.31 passing through a semi-permeable membrane, and the minimal compression work is in Eq. 2.32.

$$\frac{\dot{W}_{comp}}{n_{react}} = \sum_{react} \nu_{react} \bar{R} T_0 ln \left(\frac{P_{react}}{P_0} \right) \tag{2.32}$$

For the gases formed will be expanded from their partial pressure (P_L, P_M) to T_0 and P_0, as indicated in Eq. 2.33

$$\frac{\dot{W}_{expansion}}{n_{react}} = \sum_{prod} \nu_{prod} \bar{R} T_0 ln \frac{P_0}{P_{prod}} \tag{2.33}$$

We can use the same idea for the products; nevertheless, each component passes through a semi-permeable membrane and passes from its partial pressure in the reaction (using compressor/turbines). The maximum work that can be obtained by this process is in Eq. 2.34: $\dot{W}_{MAX} = \sum_{reactant} \dot{W}_{expansion} - \sum_{products} \dot{W}_{compression}$.

$$\frac{\dot{W}_{MAX}}{\dot{n}_T} = -\bar{R} T_0 \left(\sum_{reac} \nu_{reac} ln \frac{P_{reac}}{P_0} + \sum_{prod} \nu_{prod} ln \frac{P_0}{P_{prod}} \right) \tag{2.34}$$

For this specific case, Eq. 2.34 becomes 2.35, and K_P being the since the maximum available work does not change as a function of partial pressure modifications, being called as equilibrium constant, we can obtain Eq. 2.36.

$$\frac{\dot{W}_{MAX}}{\dot{n}_T} = \bar{R} T_0 ln \frac{P_L^{\nu_L} P_M^{\nu_M}}{P_A^{\nu_A} P_B^{\nu_B}} \times ln(P_0)^{(\nu_L + \nu_M - \nu_A - \nu_B)}) = K_P \tag{2.35}$$

Combining all effects of maximum available work and naming $W_{MAX}/\dot{n}_T = -\Delta g(T_0, P_0)$, we obtain that the equilibrium constant K_P.

$$\frac{\Delta g(T_0, P_0)}{\bar{R} T_0} = -K_P \tag{2.36}$$

Sometimes, there is a misinterpretation between the thermodynamic properties (i) exergy of a flow–$\bar{h} - \bar{h}_0 - T_0(\bar{s} - \bar{s}_0)$–with the (ii) Gibbs function $\bar{g} = \bar{h} - T\bar{s}$. Nevertheless, this last property, $\bar{g}$ is only a fundamental one, with no such interpretation of capability to perform work. Nevertheless, [5] points out that the exergy of a given flow rate is nothing more than a Gibbs free energy variation $(\Delta \bar{g})$, being one of the states the reference environment, and the energy exchange is with this environment $(\dot{Q}_0)$ at T_0 and P_0. That is, given an environment like the one proposed by [3], it is enough to calculate the difference in Gibbs free energy between the current state $(g(T, P))$ and the dead state $(g(T_0, P_{0,i}))$, showing that exergy is a thermodynamic property. Therefore, $\bar{b}(T_i, P_i) = \bar{g}(T_i, P_i) - \bar{g}(T_0, P_0)$

$$\bar{b}(T_i, P_i) = \bar{h}(T_i, P_i) - \bar{h}(T_0, P_0) - T_0(\bar{s}(T_i, P_i) - \bar{s}_0(T_0, P_0)) \tag{2.37}$$

When assessing an ideal gas (subscript: g) that is part of the reference environment, such as $g = O_2$, N_2, CO_2, with a partial pressure in this environment, ($P_{g,0}$) in this environment and the same gas at P_0, is there any available work? Applying Eq. 2.37 it is possible to evaluate the maximum amount of work available in the pressure difference between the stage change from P_0 and T_0 to $P_{g,0}$ and T_0. Therefore, for the chemical exergy, this becomes more straightforward.

$$b(T_0, P_{g,0})^{\Delta\mu} = \bar{h}(T_0, P_0) - \bar{h}(T_0, P_{g,0}) - T_0(\bar{s}(T_0, P_0) - \bar{s}(T_0, P_{g,0})) \quad (2.38)$$

Using Gibbs relations $TdS = dH - \forall dP$, for this change of state the result is $\bar{g}(T_0, P_0) - \bar{g}(T_0, P_{g,0}) = \bar{R}T_0 ln(P_0/P_{g,0})$. Hence Eq. 2.39 shows the maximum available work of a gas present in the atmosphere and its real dead state.

$$\bar{b}(T_0, P_{g,0}) = \bar{R}T_0 ln\left(\frac{P_{g,0}}{P_0}\right) \quad (2.39)$$

The interpretation of this equation is an isothermal and reversible turbine exchanging heat with the environment to take a specific gas from the restricted reference to a dead state, where there is no difference in the chemical potential between the gas and the same gas in the environment. We can understand this as there is no:

- No kinetic energy and potential difference; in other words, not a possibility to move the object or the flow without external work.
- No temperature difference with the environment (thermal exergy). There is no possibility of transferring energy as heat.
- No pressure difference between the component, in this case, gas, with the reference condition. its pressure achieved the P_0 and them $P_{g,0}$. This last movement is based on diffusion and relates to the chemical potential in a physique of random movement of each molecule that seems an organized diffusion in a larger picture, as well illustrated by [17].

Suppose we remember the example of the chocolate (let us represent it by several glucose molecules combined–$C_6H_{12}O_6$). Therefore, in the chemical bond, there is available energy to perform work by chemical reactions; hence, the glucose is not at an equilibrium state with the environment ($\mu_{gluc} \neq \mu_0$). To calculate the chemical exergy of a pure substance that is not present in the reference environment, for example, a hydrocarbon, one can use the method indicated in Fig. 2.11, pointed out by [2, 3]. Figure 2.11 represents a reversible chemical reaction that occurs at T_0 and p_0, in which the substance reacts with elements brought from the environment (co-reactants) and generates other elements that already exist in the environment (products). The second process is a change in the concentration of the co-reactants (from the state in which they are in the environment to the state in which they are pure at T_0 and P_0) and of the products (from T_0 and P_0 to the state of equilibrium with the environment).

Therefore, this process is the complete degradation of a molecule to environmental conditions, with any disequilibrium. Moreover, the idea of reversible reaction is quite

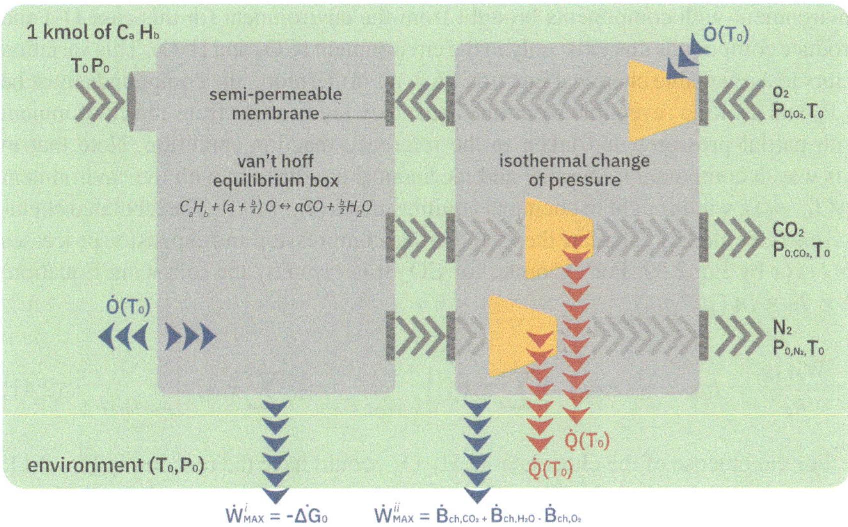

Fig. 2.11 A schematic representation of the standard chemical exergy of a hydrocarbon. The method is valid for any substance not in the reference environment. *Source* Own authorship, based on [2, 3]

unusual. Nonetheless, the Van't Hoff Reaction Box shows how this occurs, where every component is a T_0 and P_0 at the inlet and outlet. Some authors [3] and [2] show this box in the first chapter of their book. Now, the complement of exergy analysis is to use isothermal turbines and compressors to bring components that exist only in the environment, according to the definition of chemical exergy.

Equation 2.40 indicates the reversible reference reaction, in which the hydrocarbon (or another chemical compound) reacts with oxygen to assemble carbon dioxide and water. The oxygen in the medium passes through a selective permeability membrane followed by a reversible isothermal compression, as shown in Fig. 2.11. After the hydrocarbon oxidation reaction (Eq. 2.40), the products (water and carbon dioxide) will undergo a reversible isothermal expansion and then return to the environment with their respective equilibrium concentrations in the environment.

$$C_a H_b + \left(a + \frac{b}{4}\right) O_2 \leftrightarrow aCO_2 + \frac{b}{2} H_2O \tag{2.40}$$

Thus, Eq. 2.41 presents the chemical exergy of the hydrocarbon in inquiry. Note that from the moment the compound reaches the temperature and pressure of the environment, it can no longer perform mechanical work (thermal and physical). However, this chemical element is not in thermodynamic equilibrium with the environment. Such a compound must be reacted isothermally exchanging heat with the environment with components brought from the environment (in this case O_2) and produce compounds that exist only in the environment (CO_2 and H_2O). This variation refers to a reversible chemical reaction ($\Delta \bar{g}_0$). Furthermore, the compounds must be at P_0 and T_0 for a reversible reaction. Thus, they are brought from the environment with partial pressures and taken to the reference reaction condition. Note that in this way, a compound in thermal and mechanical equilibrium with the environment ($\mu(T_0, p_0)$), was brought to chemical equilibrium ($\mu(T_0, p_{g,0})$). The tabulated chemical exergies already calculate these isothermal compression and expansion processes ($\bar{b}_{ch,i}$) or by Eq. 2.39. For example, for CO_2 it is given by the following Equation: $\bar{R} \times T_0 \times ln\left(p_0/p_{CO_{2,0}}\right)$.

$$\frac{\dot{W}_{MAX}}{\dot{n}_T} = \bar{b}_{ch} = -(\Delta \bar{g}_0) - \sum x_i \bar{b}_{q,i}\Big|_{co-reactants} + \sum x_i \bar{b}_{q,i}\Big|_{products} \tag{2.41}$$

For the glucose of the chocolate–$C_6H_{12}O_6$–would have the reaction in Fig. 2.11, the reaction in the Van't Hoff reaction would have a maximum work available per kmol of $C_6H_{12}O_6$ (gluc) demonstrated at Eq. 2.42, bearing in mind that Van't Hoff equations $\dot{W}_{MAX,reac}/\dot{n}_{gluc} = -\Delta \bar{g}_{cluc}(T_0, P_0)$.

$$-\Delta \bar{g}(T_0, P_0) = \bar{g}_{gluc} + \nu_{O_2}\bar{g}_{O_2} - \nu_{CO_2}\bar{g}_{CO_2} - \nu_{H_2O}\bar{g}_{H_2O} \tag{2.42}$$

Nevertheless, we must bring the O_2 from the environment with its own partial pressure at the environment $P_{O_2,0}$ to P_0, being $\bar{b}_{ch,O_2}\bar{R}T_0 ln P_{O_2,0}/P_0$. The same reasoning can be conducted to the products, such as CO_2; nevertheless, it suffers an expansion, $\bar{b}_{ch,CO_2} = \bar{b}^{\Delta\mu} = b\bar{R}T_0 ln(P_0/P_{CO_2,0})$. Equation 2.43 shows the other term of the chemical exergy related to $\sum x_i b_{ch,i}\big|_{co-reactants} + \sum x_i b_{q,i}\big|_{products}$

$$\frac{\dot{W}_{MAX}}{\dot{n}_{gluc}} = \bar{b}^{\Delta\mu} = \bar{R}T_0\left(\nu_{CO_2} ln\frac{P_0}{P_{CO_2,0}} + \nu_{H_2O} ln\frac{P_0}{P_{H_2O,0}} + \nu_{O_2} ln\frac{P_{O_2,0}}{P_0}\right) \tag{2.43}$$

$$\frac{\dot{W}_{MAX}}{\dot{n}_{gluc}} = -\Delta \bar{g}(T_0, P_0) + \nu_{CO_2}\bar{b}_{ch,CO_2} + \nu_{H_2O}\bar{b}_{ch,H_2O} - \nu_{O_2}\bar{b}_{ch,O_2} \tag{2.44}$$

That is, $\Delta \bar{g}_0$ is the variation of the Gibbs function of the hydrocarbon oxidation reaction (with reactants and products at T_0 and P_0), $\bar{b}_{ch,i}$ is the standard chemical exergy of the element i in the reference environment and x_i, the mole fraction of element i in the reference reaction. For some more complex substances or mixtures, the calculation of chemical exergy is not trivial. Thus, chemical exergy can be calculated as a function of the lower calorific value (LHV), as indicated by Eq. 2.45. The β factor can be calculated using correlations based on fuel composition, according to [2, 3] and others.

$$\beta = \frac{b_{ch}}{LHV} \tag{2.45}$$

2.3 Some Applications of the Exergy Analysis to Usual Thermal Equipment of Daily Usage Seeking a Rational Use of Energy

In the past decades, the exergy analysis has been spreading to properly evaluate society's irreversibilities and use of natural resources works of [18–20] and several others. As discussed by [20], the neoclassical economy focuses on the allocation of resources by labeling resources, including them in black boxes of monetary systems; on the other hand, the exergy analysis "enables the analysis of resource use in a social system without subjective judgment", always involving the quality and quantity of energy. Moreover, [21] discusses that exergy shows the processes where the stable operation and evolution of the social system are maintained, suppressing the economic decision when the environment and supporting the future of the next generations are constantly examined.

Like the first law of thermodynamics, the second law of thermodynamics, employing exergy analysis, is a valuable tool for improving the efficiency of energy conversion processes and contributing to decision-making by identifying areas with higher losses of work capability or desired end product. Among others [11], the prime attempt uses the concept of lost work ($\dot{W}_{lost}$) to evaluate efficiency parameters and the quality of energy resources use. Moreover, some studies used the exergy analysis to propose a different class of energy taxation, being [22] one of the first authors in the area. Thus, in the study of energy planning, the evaluation of the life cycle of energy-intensive products and processes combined with their exergy analysis is of great interest. The cumulative exergy consumption indicator (CExC) developed by [3], and the renewability exergy index has gained attention [7]; however, there is currently no consensus in the literature regarding an exergy index. The concept of extended exergy analysis that spread in literature, proposed by [16] discusses evaluating the exergy from the cradle to the grave, similar to LCA but on an exergy basis, taking into consideration the return of the ecosystem to its original configuration. Figure 2.12 shows a schematic of the National Energy Conservation Label obtained for an air-conditioning unit in [23]. For this type of appliance, the objective is to remove energy from the environment (the evaporator) with a certain expenditure of electrical energy (in the compressor of the system). As an example, the model "split high-wall" was chosen. The classes of grade (A being the highest grade) are: (i) A, for COP higher than 3.23; (ii) B, for COP between 3.02 and 3.23; (iii) C, for values between 2.81 and 3.02; and (iv) D, for values between 2.60 and 2.81. There is a new category of COP and label in Brazil reaching values as high as 6.0. For the sake of simplicity, it shows a reasonable track of Brazilian energy efficiency policies. The complete information of the label may be seen in [24] and [25].

One point that must raise attention is the necessity of creating arguments to show that a device operates according to some quality standard. Investing in energy efficiency and conservation is essential, aiming to decrease environmental burdens due to energy generation and use. In Brazil, the main initiatives targeting consumers are the programs led by INMETRO-PROCEL (electric devices) and PETROBRAS-

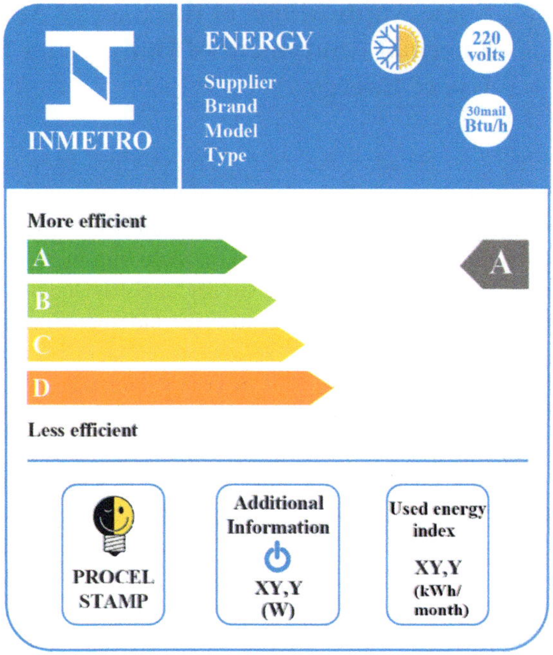

Fig. 2.12 Image of the National Energy Conservation Label, obtained for an air-conditioning unit in [23]. This figure was not translated in order to show the unmodified form of label display

CONPET (fuel-burning devices) [23]. The PROCEL index is in Fig. 2.12, since it is for an air-conditioning that corresponds to some quality standard (which evolves over time).

2.3.1 Energy and Exergy Use During Bath

Figure 2.13 exhibits three primary technology methods employed in Brazil to warm water. For the sake of comparison, energy efficiency alone may not be the best tool for differentiating these three methods since the nature of the energy inputs is different for the same product (water with a given temperature). The electric shower is in Fig. 2.13a, the gas water heater in Fig. 2.13b, and the solar water heater in Fig. 2.13c. These technologies are well documented in Brazil. Nevertheless, the electric shower is the most used due to its low costs, high energy efficiency, and ease of installation. The index employed for electric shower labeling is the power consumed, ranging in grade from A to G, since, for all circumstances, the energy efficiency (Eq. 2.46) is higher than 95%. The grade "A" was used for power consumption lower than 2400 W and "G" for power consumption higher than 7900 W [23]. For passage gas water heaters (also known as "immediate" heaters), the energy efficiency is in Eq. 2.47. The highest efficiencies are labeled "A" (84%), and the lowest are tagged as "E" (76%). For solar water heaters, the index used was the average monthly energy production per

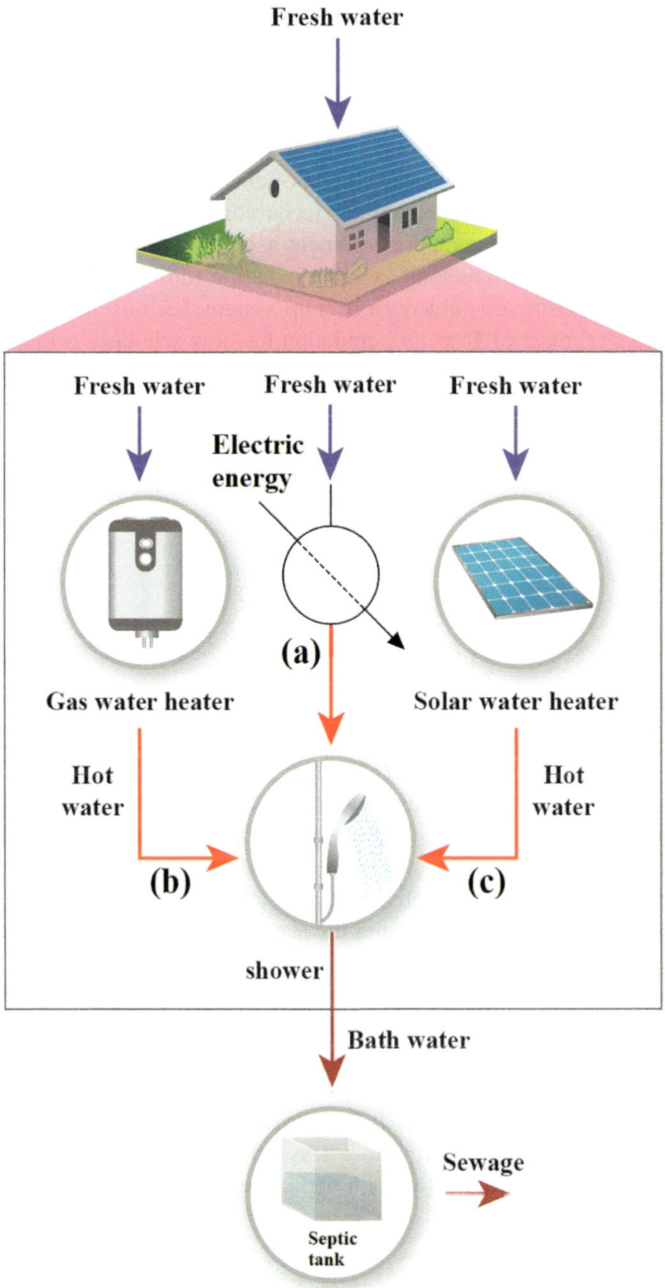

Fig. 2.13 Shower appliances: **a** electric shower; **b** gas water heater; and **c** solar water heater. *Source* [26]

square meter (PME), "A" is 80.3 kWh/(month.m^2) and "E" $52.3 < PME < 59.3$ kWh/(month.m^2). Energy efficiency does not follow the same trend because several factors affect system performance, such as the area and the material.

In contrast, the energy produced is a more helpful index since solar power is infinite from the viewpoint of humans. Nevertheless, there is no viable and ready solution regarding the different possibilities of energy use to produce similar outcomes. Hence, comparing or choosing the technology without any additional index is impossible. For solar heaters, the argument is payback time. Nevertheless, there is also no carbon footprint as long as some radiation (direct or diffuse) exists. We must not forget that there is always a backup system, such as an electrical or gas heater. Moreover, most of Brazil's population has low scholarly achievement and incomes. There must be an index that can compare these technologies without any other interpretations or calculus.

$$\eta_{eletric,shower} = \frac{\Delta \dot{H}_{water}}{\dot{W}_{input}} \tag{2.46}$$

$$\eta_{gas,heater} = \frac{\Delta \dot{H}_{water}}{\dot{m}_{gas} PCI} \tag{2.47}$$

$$\eta_{solar,heater} = \frac{\Delta \dot{H}_{water}}{\dot{I}_{solar} A_{collector}} \tag{2.48}$$

In these equations, the terms $\Delta \dot{H}_{water}$ are the enthalpy variation of the water being heated, $\dot{W}_{input}$ is the electrical energy, $\dot{m}_{gas} PCI$ is the heat of combustion of the natural gas. Also, $\dot{I}_{solar} A_{collector}$ is the solar irradiation multiplied by the area of the collector.

These previous indexes are satisfactory in helping the consumer to select the most suitable piece of equipment using the same type of equipment, e.g., compare electrical showers, as in [23]. Nevertheless, how can one choose between electrical showers and gas heaters? The price of the bills is a possibility. Regardless, geopolitical conditions may change the costs over time. How can one decide the technology that harms the environment less? Additional information could be provided in the label for comparison between technologies. Again, the idea is to facilitate the comparison. If the comparison is another index? It is necessary to use another mechanism. The exergy analysis is an analytical tool to assess the quality of the energy conversion process. The destroyed exergy for each appliance is calculated according to Eqs. 2.49–2.51.

$$\dot{B}_{d,electric} = \dot{W} - \Delta \dot{B}_{water} \tag{2.49}$$

$$\dot{B}_{d,gas} = \dot{B}_{gas} - \Delta \dot{B}_{water} \tag{2.50}$$

$$\dot{B}_{d,solar} = \dot{B}_{solar} - \Delta \dot{B}_{water} \tag{2.51}$$

In these equations, $B_{d,i}$ is the destroyed exergy of the component i, $\dot{W}$ is the power input of the electric shower, $\dot{B}_{gas}$ is the chemical exergy of the natural gas [3], $\dot{B}_{solar}$ is the exergy associated with the solar radiation [27], and $\Delta\dot{B}_{water}$ is the exergy variation of the water.

Equations 2.52–2.54 show the exergy efficiency of the appliances given in Fig. 2.13, where $\eta_{ex,i}$ is the exergy efficiency of the appliance i. These equations show a rational tool based solely on physics, which may address the problem of comparing different technologies on the same thermodynamic basis (meaning their work performance capability). Hence, future works could further step in the rational use of energy since the comparison is entirely objective. The ratio of the two terms shows the quality of the energy source. By using the extended exergy analysis, it is possible to make this index suitable for several parts of the country with different necessities. For solar heaters, it was considered that the average solar radiation for an inclined plane was $I = 0.7$ kW/m^2 [28] and the sun temperature (T_{sun}) was 5777 K.

$$\eta_{ex,eletric} = \frac{\Delta\dot{B}_{water}}{\dot{W}_{input}} \tag{2.52}$$

$$\eta_{ex,gas} = \frac{\Delta\dot{B}_{water}}{\dot{B}_{gas}} \tag{2.53}$$

$$\eta_{ex,solar} = \frac{\Delta\dot{B}_{water}}{\dot{B}_{solar,radiation}} \tag{2.54}$$

From the tables of energy consumption and efficiency [23], it was determined that the best, first law, efficiencies of electric showers exceeded 95%, the ones of instant gas heaters approached 84%, and the ones of solar collectors achieved approximately 70%. It is shown from calculations in Table 2.2 that, by the nature of energy conversion for water heating, all types of equipment were found to have a low exergy efficiency. Moreover, this physical quantity was very similar among all equipments, nearly 3%. Therefore, in all circumstances, 97% of the amount of exergy was destroyed to heat water for a shower. For this particular appliance, the second law efficiency was inexpressive. Therefore, it would be required to further assess the electricity generation efficiency. For example, it would be helpful to determine if it is renewable or non-renewable, proximal or remote from the urban center, or if it has any other distinguishing characteristics [29, 30]. From these references, it was found that the energy and exergy efficiency are both needed in order to generate each kW of electricity consumed by the electric shower and for gas flaring (in gas heaters). It is at this stage that solar heaters are distinguished because their input of exergy is almost entirely from renewable sources, as highlighted in [18].

Table 2.2 demonstrates that solar heating exergy efficiency may be considered the best technology to produce bath water. This evaluation takes into account that the ceilings of buildings are generally not used for any other purpose. The exergy efficiency of solar heating is similar to the other two technologies, while the energy

Table 2.2 Comparative energy and exergy analysis of studies of the three technologies: electric shower, gas water heater and solar heater; for a temperature increase in the system of 20 °C from the environmental conditions $T_0 = 25$ °C; in this table, only the best pieces of equipment were analyzed

Equipment	Variable	Energy index	Equation	η_{ex} (%)
Electric shower	Power	7 kW ($\eta_{en} > 95\%$)	(2.52)	2.6
Gas water heater	η_{en}	84%	(2.53)	2.6
Solar heater	PME	80.3 kWh/(month.m^2)	(2.54)	2.7

efficiency of this appliance is lower than the others. However, because the exergy input is solar, it may be considered as an "infinite renewable source". The discussion over which type of solar heating may be more efficient, for instance, a comparison between photovoltaic panels and solar water heaters, is out of the scope of this article. As discussed in Pinto and Mady [13], the photovoltaic panels may be more suitable, whereas if there is cogeneration, the solar water heaters may lead to better efficiency (for the same area).

In 2020, the share of renewable energy sources in electricity generation was 84.8% in Brazil, whereas its stake in the domestic energy supply (energy matrix) was 48.4%. There is a high percentage of the electrical matrix based on hydroelectric power, about 62%. Considering all renewable sources in electrical energy, the renewable share is responsible for 67% (where solar and eolic are increasing over time), followed by natural gas with a share of 13% [31]. The Brazilian residential sector was accountable for 10.8% of the country's final energy consumption. The primary energy source is electricity, followed by firewood; however, natural gas (and GLP) is presenting an increase in its total amounts of consumption, leading to a question of energy adding or transition [32]. Any improvement in energy use may lead to significant gains for society, although as described by [18], there is always a rebound effect when the appliance leads to a decrease in energy consumption.

Taking into account the production path of electrical energy, the data proposed by [30] on the efficiency of energy-to-electricity conversions such as hydroelectric, 82%, and cogeneration with natural gas, 45%; it is possible to assess a more comprehensive indicator for these appliances. The exergy efficiency of the electric shower becomes 2.1% (hydroelectric) and 1.2% (cogeneration with natural gas). Additionally, knowing that the hydroelectric barrier is generally far from the consumer center, transmission wastes can be considered as 16%, according to [30]. One can correct the values of Table 2.2. This fact may lead to the conclusion that if there is energy conversion based on natural gas (or coal, oil, and non-renewables) as an electrical energy source, one must consider other applications than electrical power for a shower and other appliances that have the objective to increase the temperature of the water.

2.3.2 Energy and Exergy Analysis of Air-Conditioning

Another benchmark of the necessity of the second law of thermodynamics is comparing two different types of refrigeration cycles. Figure 2.14 shows a vapor compression system (a) and an absorption cycle (b). The prime is in appliances of any size, from small-scale residential (split high-wall) to commercial and industrial applications. The absorption refrigeration cycle is usually for industrial applications, where it is necessary to dry air using residual steam as studied by [33], for commercial purposes. One interesting application is to use a lower amount of energy in trucks using as input the flue gases [34] aiming at decreasing the carbon footprint in the transportation sector in Brazil and other countries. However, the increased use of solar thermal heaters may also make possible the use of absorption cycles in residential

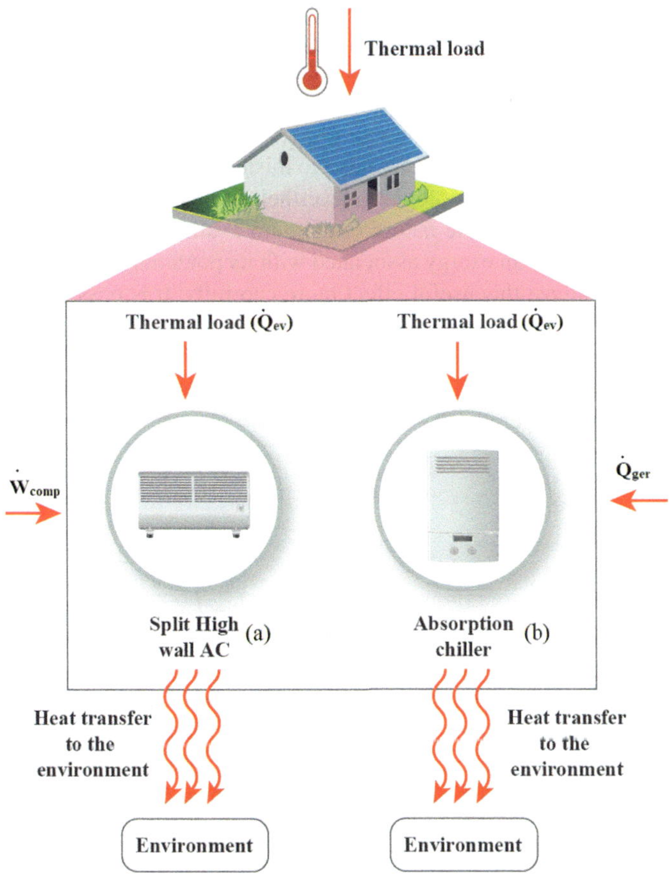

Fig. 2.14 **a** Vapor-compression refrigeration cycle; **b** absorption cycle using the mixture of water and lithium bromide. *Source* [26]

systems [35]. Nevertheless, there must be an improvement in the application of technology to avoid traditional problems, e.g., the incrustation of salt or ammonia leaking and their oversize.

For this investigation, both cycles were simulated based on the methods of [36]. The mass, species, and energy balance solution was necessary to determine the thermodynamic states and generate metric indexes. Fixing the same amount of power removed from the environment, $\dot{Q}_{evap} = 12,000$ Btu/h, gave the input variable for the cycle. Although this value was low for absorption cycles, it represents a theoretical basis of comparison to demonstrate the applications of the exergy efficiency.

In Fig. 2.14a, $\dot{W}$ is the power delivered to the cycle, and $\dot{Q}_{ev}$ is the heat removed from the environment in the evaporation process. For Fig. 2.14b, $\dot{Q}_{ge}$, is the energy transferred in the generator to the absorption cycle.

For the vapor-compression refrigeration cycle (Fig. 2.14a), the coefficient of performance (COP) and the exergy efficiency are evaluated according to Eqs. 2.55 and 2.56. It was possible to infer from these equations that both can be used to differentiate between the devices. However, Eq. 2.56 gives the highest value of 100% (where the efficiency would be the Carnot's), whereas Eq. 2.55 can achieve figures higher than unity (being the maximum value the Carnot's), considering that it compares heat rate removed from the environment, $\dot{Q}_{evap}$ (lower quality) with electric power, $\dot{W}_{comp}$ (higher quality). This fact may direct to misleading information when there is a comparison of different technologies. Nevertheless, since the exergy analysis uses physical quantities of the same nature (maximum available power), e.g., in a heat transfer, there is always an exergy associated with its potential to perform any work. Moreover, let's analyze this useful effect more carefully. It is possible to conclude that while heat is removed from the environment from the evaporator, the exergy is transferred to the environment (increasing the disequilibrium with the reference ambient) as discussed in [2]. The air conditioning increases the exergy to the thermal environment to be refrigerated.

$$COP = \frac{\dot{Q}_{evap}}{\dot{W}_{comp}} \tag{2.55}$$

$$\eta_{ex,comp} = \frac{\dot{Q}_{evap}\left(1 - \frac{T_0}{T_{evap}}\right)}{\dot{W}_{comp}} \tag{2.56}$$

The absorption conditioning systems (usually available as a chiller), Fig. 2.14b, can be evaluated according to its COP (Eq. 2.57). This value is usually lower than the related to vapor compression cycles, since this type of chiller compares the energy of the same quality (heat transfer rate). It compares the heat provided in the generator with the heat removed from the environment (usually $\dot{W}_{pump}$ is smaller than other types of energy transfer). Therefore it is almost **impossible** to compare both cycles only using the COP as a metric.

$$COP = \frac{\dot{Q}_{evap}}{\dot{Q}_{gen} + \dot{W}_{pump}} \tag{2.57}$$

$$\eta_{ex,abs} = \frac{\dot{Q}_{evap}\left(1 - \frac{T_0}{T_{evap}}\right)}{\dot{Q}_{gen}\left(1 - \frac{T_0}{T_{gen}}\right) + \dot{W}_{pump}} \tag{2.58}$$

Equations 2.56 and 2.58 describe the ratio of the same quality (exergy). In the denominator, the exergy of the heat in the generator, $\dot{B}_{Q_{gen}} = \dot{Q}_{gen}\left(1 - \frac{T_0}{T_{gen}}\right)$, and the power delivered in the compressor ($\dot{W}_{comp}$). In the numerator of both equations, there is the exergy transfer associated with the evaporation process, $\dot{B}_{Q_{evap}} = \dot{Q}_{evap}\left(1 - \frac{T_0}{T_{evap}}\right)$. With exergy efficiency, it was possible to compare two distinct technologies. This exergy efficiency is a complementary index to the COP, to help the customers decide which technology to invest in. It is necessary to bear in mind that no absorption chiller is available as easily as a split high wall. The objective of this example is a proof of concept. It was adopted for the two systems (i) evaporation temperature (T_{ev}) of 3 °C; (ii) rejected heat to the environment at 25 °C (T_0); (iii) generator temperature (T_{ge}) at 80 °C.

$$\dot{B}_{d,comp} = \dot{W}_{comp} - \dot{Q}_{evap}\left(1 - \frac{T_0}{T_{evap}}\right) \tag{2.59}$$

$$\dot{B}_{d,abs} = \dot{Q}_{gen}\left(1 - \frac{T_0}{T_{gen}}\right) + \dot{W}_{pump} - \dot{Q}_{evap}\left(1 - \frac{T_0}{T_{evap}}\right) \tag{2.60}$$

Table 2.3 indicates the performance indexes of two conditioning systems: vapor compression and absorption. For the vapor compression system, the fluid R410A was chosen, with the specifications found in general technical catalogs. The model was adapted from [36] for the absorption system.

The general energy metric indexes, COP, are nearly 3.0 for compression vapor and 0.8 for absorption chillers. The aim of this index in the prime cycle is to inform the amount of energy removed from the environment ($\dot{Q}_{evap}$) to the power input ($\dot{W}_{comp}$). In contrast, the absorption system is the ratio of the $\dot{Q}_{evap}$ to the $\dot{Q}_{gen}$ ($\dot{W}_{pump}$ is negligible). The nature of the denominators of these equations is different based on the following explanation: the first denominator is electrical power, and

Table 2.3 Comparative energy and exergy analyses of conditioning systems, for fluid temperature leaving the condenser at 45 °C

–	Eq.	En.$_{Metrics}$ (-)	Eq.	Ex.$_{Metrics}$ (%)
Vapor comp. system	(2.55)	$COP = 3.1$	(2.56)	$\eta_{ex} = 27.3$
Absorption system	(2.57)	$COP = 0.75$	(2.58)	$\eta_{ex} = 33.5$

the second is heat transfer rate. It is logical that the indexes should not be compared because these are two distinctly different cycles. When the exergy analysis is applied to these systems, the absorption cycle yields higher figures; therefore, this cycle uses the exergy provided more efficiently. It is essential to highlight that the results of Table 2.3 are restricted to one condition ($Q_{evap} = 12{,}000$ btu/h), which may be extrapolated to several thermal loads; nevertheless, the method to compare both cycles is the same and is practical for residential, commercial, and industrial applications, as in [33] and [34].

2.3.3 *Internal Combustion Engine*

The internal combustion engine (ICE) is currently the dominant power source for vehicles, ICEV (light and heavy duty); the following paragraphs will show the exergy content of each stream available to obtain a better thermal integration of ICEV. The pursuit of efficiency gains can often come at a high cost for relatively small improvements. Figure 2.15 illustrates the control volume used in the analyses conducted in this study. Rovai and Mady [37] focused on a spark ignition, four-stroke, flex-fuel, naturally aspirated, multipoint indirect injection engine tested on a dynamometer bench under steady-state conditions. The engine tests occurred using two boundary fuels: E22, which is gasoline sold in Brazil, and E100 (hydrous ethanol, that is, a content of 94% and 6% of water), under the same speed and load conditions.

Figure 2.16 presents a Sankey diagram of an engine running on E22 and E100 fuels. The diagram shows that the efficiency of the engine using ethanol is slightly higher than that of the engine compared to gasoline. What is interesting in this figure is that the energy transfer to the cooling fluid is about 33% of the overall energy input. The energy content of the flue gases has similar figures. The rest of the energy transferred to the environment is about 1% of the total fuel's energy. It is impossible to assess further information regarding the potential utilization of this energy form using only the first law of thermodynamics. One can raise the question of whether applying the first law can answer whether it is possible to use about 2/3 of the energy input of the fuel ($\dot{Q}_{fuel} = \dot{m}_{fuel} LHV_{fuel}$) for other processes.

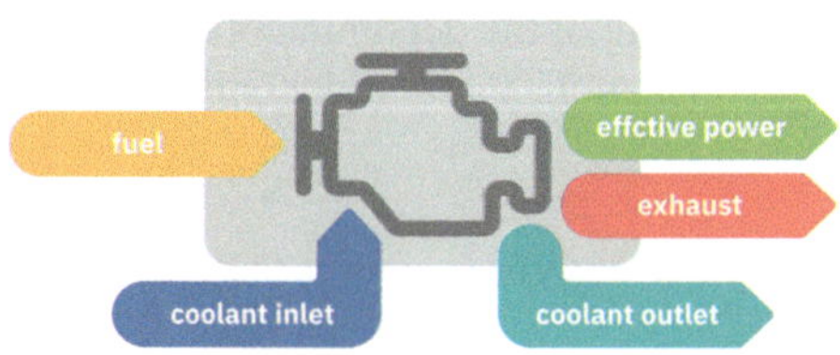

Fig. 2.15 Engine control volume, with fuel intake, coolant fluid input and output, heat transfer to environment, and performed work. *Source* [37]

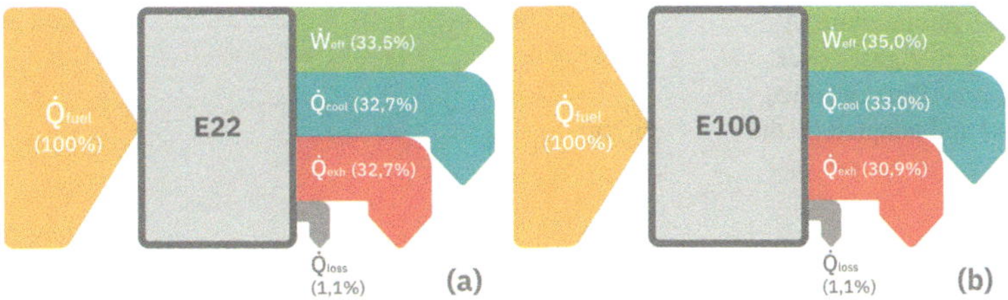

Fig. 2.16 Sankey energy diagram for E22 **a** and E100 **b** with useful work and percentage of heat transfer lost to the environment. *Source* Modified and adapted from [37]

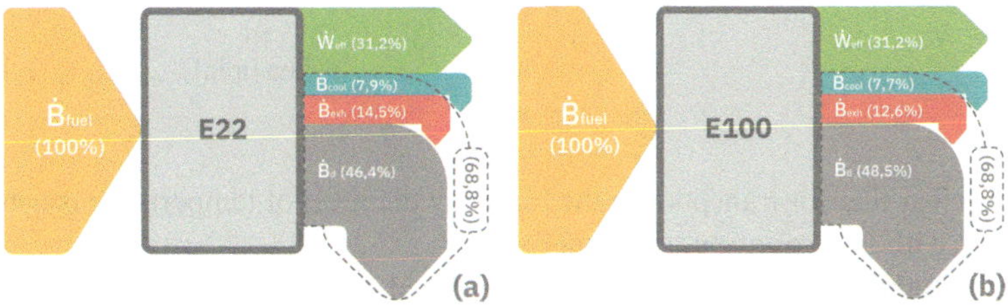

Fig. 2.17 Grassmann exergy diagram for E22 **a** and E100 **b** with useful work and percentage of exergy transfer associated with heat transfer lost to the environment. *Source* Modified and adapted from [37]

Now, when the exergy analysis is applied, it becomes possible to understand these energy transfers' quality of work potential. In Fig. 2.17, it is clear that 50% of the exergy content of the fuel in the engine is destroyed by internal irreversibilities, already showing some additional information. However, this tool reveals that the cooling fluid has the potential to perform work equivalent to approximately 8% of the chemical exergy of E100 or E20, although energy analysis answers that 33% is available for use without taking into consideration the quality of the energy. Moreover, the exergy content of the flue gas represents 12 to 14% of the chemical exergy of the fuel. This finding suggests that utilizing the flue gas to power an absorption chiller could be a viable option to increase engine efficiency. As proposed by [34], this replacement could reduce the power consumption with compression vapor systems connected to the crankshaft and increase the vehicle's engine efficiency by better thermal integration.

This idea is linked with a work proposed by [34], where the substitution of the compressor directly activated by the crankshaft system would have a replacement by the use of energy content in flue gas at high temperatures at the outlet of the engine. Examples are minibuses, buses, trunks, and mostly refrigeration trucks. The experimental tests were carried out in June 2017 in São Bernardo do Campo, São Paulo, Brazil. The average ambient temperature was 18.5 °C. Figure 2.18 shows the temperature of the flue gases on a highway and in the city (warm-up and complete cycle). These levels of temperature are usual for the absorption refrigeration cycle.

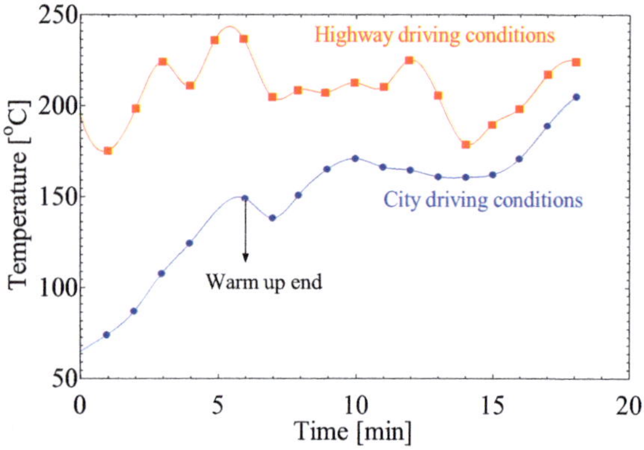

Fig. 2.18 Temperature of the exhaust flue gases over time. *Source* Data obtained from [34] and adapted by the author

And it is clear that for transport on the highway the level of temperature is achieved faster and with low variation over time.

In the model of Ranieri et al. [34], considered a fraction of the engine cooling fluid and exhaust gas for the generator's energy source in an absorption refrigeration cycle. Hence, we included recuperative heat exchangers with an auxiliary water circuit between the generator and exhaust system. The results showed an average cooling capacity of 1.5 kW during the initial warm-up phase, reaching 7.2 kW under road-route-driving conditions. This condition is the most used during its usage in the cities, where the cooling capacity is 3.6 kW, 68% of the cooling demand in vehicles in this category. Figure 2.19 shows the experimental results and the idea of changing a concept in the transportation sector, with the need for decarbonization with a higher integration of the internal combustion engine thermal system. The blue line shows the Carnot efficiency of the engine using only the flue gases during highway driving, and the red lines show the exergy efficiency of the refrigeration system [34].

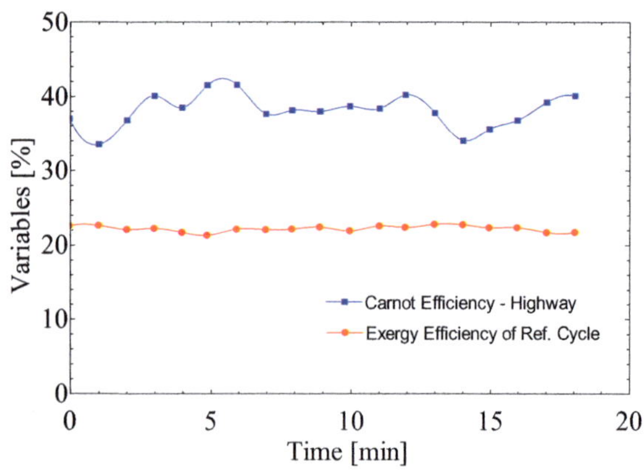

Fig. 2.19 Average (points) of Carnot efficiency **a** and exergy efficiency of the absorption cycle **b** of the exhaust system gas for the conditions studied. *Source* Adapted and modified from [34]

References

1. Prigogine I, Stengers I (1997) The end of certainty. Simon and Schuster
2. Kotas TJ (1985) The exergy method of thermal plant analysis. Butterworth Publishers, Stoneham, MA, London
3. Szargut J, Morris DR, Steward FR (1988) Exergy analysis of thermal, chemical, and metallurgical processes. Hemisphere Publishing Corporation, New York
4. Balmer RT (1982) Entropy and aging in biological systems. Chem Eng Commun 17:171–181
5. Gaggioli RA (1980) Thermodynamics: second law analysis. American Chemical Society
6. Bejan A (2016) Advanced engineering thermodynamics. John Wiley & Sons
7. Oliveira-Junior S (2021) Exergy: production, cost and renewability. Springer Science & Business Media
8. Rant Z (1955) Energy value and pricing. Strojniski Vestnik 1:4–8
9. Tester JW, Modell M (1997) Thermodynamics and its applications. Prentice Hall PTR, New Jersey
10. Gallo WLR, Milanez LF (1990) Choice of a reference state for exergetic analysis. Energy 15:113–121
11. De Nevers N, Seader JD (1980) Lost work: A measure of thermodynamic efficiency. Energy 5:757–769
12. Pereira MTRM, Carvalho M, Mady CEK (2022) Adressing energy demand and climate change through the second law of thermodynamics and lca towards a rational use of energy in brazilian households. Entropy 24:1524
13. Pinto CR, Mady CEK (2020) Comparing the thermodynamic performance of organic Rankine and Kalina cycles in solar energy systems. J Brazilian Soc Mech Sci Eng 11:597–610
14. Valero VC, Valero AV (2015) Thanatia: the destiny of the Earth's mineral resources: a thermodynamic cradle-to-cradle assessment. 1st edn. World Scientific
15. Dunbar WR, Lior N, Gaggioii RA (1992) The component equations of energy and exergy. J Energy Res Technol 114:75–83
16. Sciubba E (2001) Beyond thermoeconomics? The concept of extended exergy accounting and its application to the analysis and design of thermal systems. Exergy Int J 1:68–84
17. Schrödinger E (1944) What is life?: the physical aspect of the living cell; with. Cambridge University Press, Cambridge, Mind and matter; Autobiographical sketches
18. Mosquim RF, Oliveira-Junior S, Mady CEK (2018) Modelling the exergy behavior of São Paulo State in Brazil. J Clean Prod 197:643–655
19. Chowdhury H, Chowdhury T, Thirugnanasambandam M, Farhan M, Ahamed JU, Saidur R, Sait SM (2019) A study on exergetic efficiency vis-à-vis sustainability of industrial sector in Bangladesh. J Clean Product 231:297–306
20. Meng Z, Yang Z, Zhang B, Wu X (2023) Resource exergy analysis of the Chinese society in 2017. Environ Sci Poll Res 30(14):42016–42033
21. Hossain S, Chowdhury H, Chowdhury T, Ahamed JU, Saidur R, Sait SM, Rosen MA (2020) Energy, exergy and sustainability analyses of Bangladesh's power generation sector. Energy Rep 6:868–878
22. Hirs G (1993) Exergy loss: a basis for energy taxing. Energy efficiency in process technology. Springer, Netherlands, Dordrecht, pp 1241–1253
23. INMETRO (2013) ENCE—Etiqueta Nacional de Conservação de Energia: Selo PROCEL de Economia de Energia/ANEXO DA PORTARIA INMETRO No 410/2013. http://www.inmetro.gov.br/legislacao/
24. Vendrusculo EA, Queiroz GDC, Jannuzzi GDM, Silva-Júnior HX, Pomilio JA (2009) Life cycle cost analysis of energy efficiency design options for refrigerators in Brazil. Energy Efficiency 2:271–286
25. Cardoso RB, Horta-Nogueira LA, Souza EP, Haddad J (2012) An assessment of energy benefits of efficient household air-conditioners in Brazil. Energy Efficiency 5:433–446
26. Mady CEK, Pinto CR, Pereira MTRM (2020) Application of the second law of thermodynamics in brazilian residential appliances towards a rational use of energy. Entropy 22(6):616

27. Petela R (2003) Exergy of undiluted thermal radiation. Solar Energy 74(6):469–488
28. Ge Z, Wang H, Wang H, Zhang S, Guan X (2014) Exergy analysis of flat plate solar collectors. Entropy 16(5):2549–2567
29. Mady CEK, Bresciani AE, Paiva JL, Jardin-Junior RN, Guardani R, Oliveira-Junior S (2018) Thermodynamic assessment of MVR implementation in multistage evaporation plants. J Brazilian Soc Mech Sci Eng 40(8):395
30. Flórez-Orrego D, Silva JA, Oliveira-Junior S (2014) Renewable and non-renewable exergy cost and specific CO2 emission of electricity generation: The Brazilian case. Energy Conv Manage 85:619–629
31. EPE (2019) Enterprise of energy balance brazil. Brazilian Energy Balance 2020. Brazilian ministry of energy and mines
32. York R, Bell SE (2019) Energy transitions or additions?: Why a transition from fossil fuels requires more than the growth of renewable energy. Energy Res Soc Sci 51:40–43
33. Lima LS, Mady CEK (2021) Energy and exergy analysis of an absorption and mechanical system for a dehumidification unit in a gelatin factory. Entropy 23(4):415
34. Ranieri MA, Manieri G, Albuquerque Mady CEKC (2021) Could an absorption refrigeration system be driven by the engine exhaust gas and cooling fluid of a minibus? J Brazilian Soc Mech Sci Eng 43(12):544
35. Asghari M, Brouwer J (2019) Integration of a solid oxide fuel cell with an organic rankine cycle and absorption chiller for dynamic generation of power and cooling for a residential application. Fuel Cells 19(4):361–373
36. Herold KE, Radermacher R, Klein SA (2016) Absorption chillers and heat pumps. CRC Press
37. Rovai FF, Mady CEK (2022) Exergetic analysis of an internal combustion engine running on E22 and E100. Revista de Engenharia Térmica 21(2):43–50

Chapter 3
Exergy Analysis of the Human Body

3.1 First Law of Thermodynamics and the Human Body

Figure 3.1 indicates a representation of the human body model, based on [1], with respective heat and enthalpy transfers to the environment, food and water intake, and urine and feces outflow. The human body has clear divisions (skin) between the core and environment into two control volumes: CV1 and CV2. The first control volume (CV1) comprises the thermal, respiratory, circulatory, and other organs forming individual systems. From CV1, there are heat and enthalpy transfer rates to the environment associated with radiation ($\dot{Q}_r$), convection ($\dot{Q}_c$), vaporization ($\dot{H}_e$) and breathing ($\Delta \dot{H}_{res} = \dot{H}_{ex} - \dot{H}_a$) and also internal transfer of the heat released due to metabolism ($\dot{Q}_M$). The second control volume (CV2) represents the cellular metabolism in which the input is nutrients (glucose, lipids, and amino acids) and oxygen, and the output of water, urea, carbon dioxide, and the heat released by the reactions $\dot{Q}_M$ to CV1. The division into CV1 and CV2 is an idealization for didactic purposes and aims to facilitate modeling and make it possible to divide the body into thermal and chemical effects. In this way, it was possible to compare the changes made to the model concerning the literature that only considers the thermal portion and a detailed analysis of the metabolism. This distinguishing of the metabolism is a first step to studying cell (for cancer diagnosis or identification).

This figure and equations originate from the results of Mady and Oliveira-Junior [1] and Mady et al. [1]. Other works followed these equations as in [3]. These works followed the pioneer one of Batato et al. [4]. The works proposed by [5] and [6] brought some modifications to the model of Fig. 3.1; these modifications refer to the respiratory system of [7] and the transport of gases through the blood, as will be seen in Chap. 7.

Now, in Eq. 3.1, the total change in the internal energy of the body is indicated, which is a sum of the change in internal metabolic energy rate ($\dot{M}$) with the change in internal energy of the body due to temperature variations over time ($dU_{\Delta T}/dt$). The same can be accomplished for the temporal variation of exergy, as indicated in

C. E. Keutenedjian Mady, *Exergy Analysis*, Green Energy and Technology,
https://doi.org/10.1007/978-3-032-01646-1_3

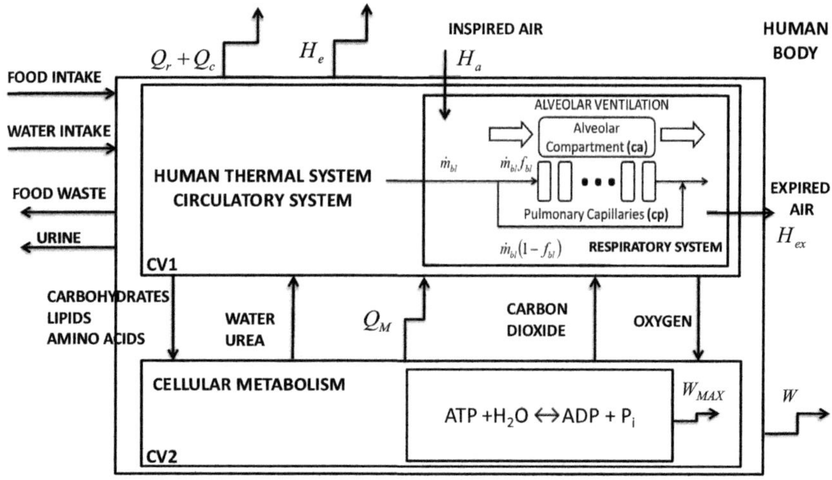

Fig. 3.1 Schematic representation of the human body, with the intake of food, water and inspired air; and output of food, urine, expired air, vaporization trough skin and heat release due to radiation and convection. *Source* Own authorship, based on [1]

Eq. 3.2, where $\mathbf{B}_{CV}$ is the body's exergy in [J] and $\dot{B}_M$ is the metabolism on an exergy basis [W].

$$\frac{dU_{Body}}{dt} = -\dot{M} + \left.\frac{dU}{dt}\right|_{\Delta T} \tag{3.1}$$

$$\frac{d\mathbf{B}_{Body}}{dt} = -\dot{B}_M + \left.\frac{d\mathbf{B}}{dt}\right|_{\Delta T} = -\dot{B}_M + \left(\left.\frac{dU_{Body}}{dt}\right|_{\Delta T} - T_0 \left.\frac{dS_{Body}}{dt}\right|_{\Delta T}\right) \tag{3.2}$$

Metabolism is defined as the enthalpy variation in the reaction of oxidation of the nutrients digested by the gastrointestinal system: $\dot{M} = \dot{H}_{reac} - \dot{H}_{prod}$, where $\dot{H}_{reac}$ is the enthalpy of the reactants (oxygen, carbohydrates, lipids, and amino acids), $\dot{H}_{prod}$ is the enthalpy of the products of cellular reactions (carbon dioxide, liquid water, and urea). From an energy balance in cellular metabolism (CV2), it can be seen that $\dot{Q}_M = \dot{M} - \dot{W}$, that is, $\dot{Q}_M = \dot{M}$ only for basal conditions, exclusively. The exergy metabolism B_M is defined as: $\dot{B}_M = \dot{B}_{reac} - \dot{B}_{prod}$, where $\dot{B}_{reac}$ is the exergy of the reactants and $\dot{B}_{prod}$ is the exergy of the products. As carried out by [1], the consideration that separating nutrients does not lead to irreversibilities is necessary. However, it is a strong hypothesis. Future works focused on the digestive tract may help to answer these doubts since there are several chemical reactions, diffusion, and other molecular characteristics.

Figure 3.2 indicates how the energy conversion process occurs in cellular metabolism from the first law perspective. Since no energy is lost, part of the metabolism becomes heat and another work. For the second law perspective, the

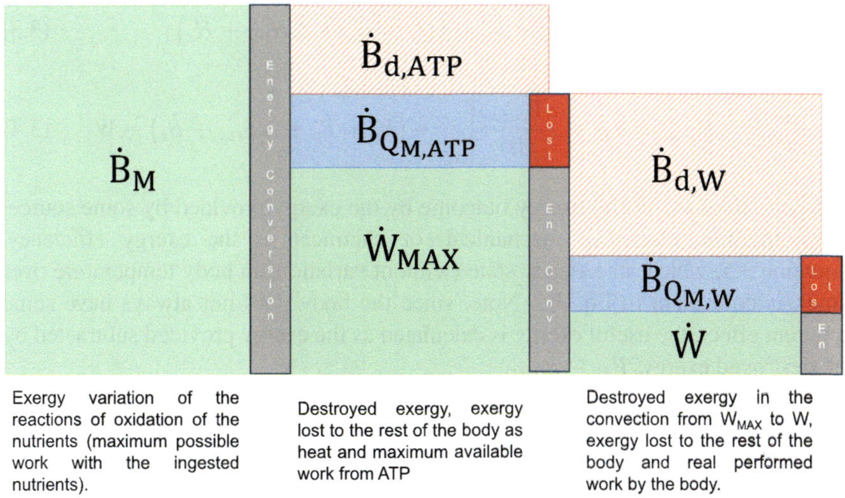

Fig. 3.2 The degradation of nutrients in the body, divided into three fundamental steps: oxidation of nutrients, formation of ATP, and use of ATP. From the first law perspective, the conversion to the ATP is not marked. *Source* [8] and [1]

exergy of nutrients is used to produce ATP (adenosine triphosphate), a substance known as "energy currency" for any energy conversion process in the body. The change in Gibbs free energy in the hydrolysis of ATP is known as the maximum theoretical work that the body can accomplish: $\Delta \dot{G}_{ATP} = \dot{W}_{MAX}$, whereas from the first law perspective, it is not possible to state which part is lost as heat or used is available for a useful effect. The second stage is transforming energy in ATP into work or vital functions. Note that the term $\dot{B}_{Q_M}$ is the exergy referring to the heat released in the metabolism $\dot{Q}_M$, calculated as $\dot{B}_{Q_M} = \dot{Q}_M \left(1 - \frac{T_0}{T_{body}}\right)$. One must take into consideration a careful analysis to consider this exergy as the body's metabolic exergy, as it is an order of magnitude smaller than the metabolism, even presenting values lower than the work performed, as indicated, respectively, in [2] and [1].

It highlights the importance of applying the first law of thermodynamics to obtain the thermal and phenomenological behavior of the body, e.g., the models proposed by [9]. It then becomes possible to obtain the energy and exergy analyses for the body, indicated by Eqs. 3.3 and 3.4. From Eq. 3.4, it becomes clear that metabolic exergy is a physiological parameter that defines the maximum work possible to obtain from the oxidation of the nutrients ingested; the irreversibilities of this reaction come from the conversion of macronutrients into the exergy content in the molecule of ATP. The rest of the irreversibilities is how good or bad the body uses this exergy for daily activities and tasks. That is the reason for $\dot{M} = \dot{Q}_M$ during basal conditions, based on experimental results of direct and indirect calorimetry since the works of [10].

$$\left.\frac{dU}{dt}\right|_{\Delta T} = \dot{M} - \dot{W} - \left(\dot{Q}_r + \dot{Q}_c + \Delta\dot{H}_{res} + \dot{H}_e\right) \tag{3.3}$$

$$\dot{B}_{dest}^{body} = \dot{B}_{dest} = \dot{B}_M - \left.\frac{d\mathbf{B}_{corpo}}{dt}\right|_{\Delta T} - \left(\dot{B}_r + \dot{B}_c + \Delta\dot{B}_{res} + \dot{B}_e\right) - \dot{W} \tag{3.4}$$

The ratio of the useful exergy outcome by the exergy provided by some source, being thermal, electrical, mechanical, or chemical, is the exergy efficiency. Equation 3.5, which, in a steady state (without variations in body temperature over time), is equivalent to Eq. 3.6. Note, since the body does not always have some apparent effect, the useful exergy is calculated as the exergy provided subtracted by the destroyed exergy, $\dot{B}_M - \dot{B}_{dest}$.

$$\eta_b = \frac{\left|\frac{d\mathbf{B}}{dt}\right| - \dot{B}_{dest}}{\left|\frac{d\mathbf{B}}{dt}\right|} \tag{3.5}$$

$$\eta_b = \frac{\dot{B}_M - \dot{B}_{dest}}{\dot{B}_M} \tag{3.6}$$

3.2 Energy and Exergy Transfer to the Environment

Every energy transfer has the potential to perform an amount of work. Therefore, each energy interaction has its exergy interaction with the surroundings, the whole body, the skin, and the environment. This section will analyze the phenomenological models of the human body with the energy transfer to the environment associated with heat (radiation and convection) and enthalpy (sweat, diffusion, and respiration) to assess the boundary conditions and the complete equation of conservation: mass, species, and energy [1–3, 9, 11–15].

3.2.1 *Energy Transfers to the Environment*

The calculus of heat and enthalpy exchanges through convection, radiation, evaporation, and respiration is necessary to apply the energy balance to the human body. It is possible to calculate the heat transfer rate by convection (Eq. 3.7) and by radiation (Eq. 3.8). In the flow rate analysis, one may consider an additional resistance to the skin's surface for the clothing according to [14] standard. Calculating heat and enthalpy transfers is necessary to use some information gathered in literature [14] and [15].

$$\dot{Q}_c = f_{cl} h_c A_d (T_{cl} - T_a) \tag{3.7}$$

Where, $\dot{Q}_c$ is the heat transfer rate by convection [W]; f_{cl} is the ratio between the area of the clothed and naked body; h_c is the convection heat transfer coefficient [W/m^2.^{o}C]; T_{cl} is the surface temperature of the clothing [°C]; T_a is the environmental air temperature [°C]; and A_d is the surface area of the body [m^2] (calculated from Du Bois [16] equation).

$$\dot{Q}_r = f_{cl} h_r A_d (T_{cl} - T_{rm}) \tag{3.8}$$

Being $\dot{Q}_{rad}$ the heat transfer rate associated with radiation (W); h_r the heat transfer coefficient by radiation [W/m^2] and T_{rm} average heat transfer by radiation [^{o}C]. Remembering that $h_r = \sigma f_r \epsilon A_d (T_{cl} + T_{rm})(T_{cl}^2 + T_{rm}^2)$. Being ϵ the emissivity of the skin (0.95); σ a Stefan-Boltzmann constant [W/m^{2}K^4]; and f_r the ratio between the effective and the external body area.

In Eqs. 3.7 and 3.8, clothing can be considered an additional resistance to heat flow (facilitating the modeling of heat transfer between skin and clothing). Thus, by applying the energy balance around the clothes, it is possible to reach Eq. 3.9.

$$\dot{Q}_r + \dot{Q}_c = \frac{T_p - T_o}{R_{cl} + \frac{1}{f_{cl} h}} \tag{3.9}$$

Where, R_{cl} is the clothing's thermal resistance to conduction; T_p, the skin surface temperature; T_o, the operating temperature; h, the combined heat transfer coefficient [W/(m^2.°C)]. Remembering that $T_o = (h_c T_a + h_r T_{rm})/(h_c + h_r)$.

It is essential to highlight that in the thermal environment, there is a composition of several phenomena, e.g., air velocity, relative humidity, the temperature of the air, and mean radiant temperature. With the temperature of the walls equal to that of the ambient air, the operating temperature will become equal to the temperature of the environmental air. The radiative and convective heat transfer coefficients were obtained from experimental work on a female mannequin, as indicated by [17], the value of the combined coefficient being the sum of the radiative and convective coefficients.

The enthalpy transfer rate as a result of evaporation on the skin can have two components: one due to the diffusion of water across the surface of the skin (as if 6% of the surface were wet), and the other, subject to the body's thermoregulatory system, caused by sweat evaporation and increased surface wetness ($d\omega$). Equation 3.10 shows the energy transfer related to the outcome of the evaporation on the skin surface. It is the water flow rate multiplied by the energy required for phase change at the skin temperature or using the diffusion mass transfer coefficient.

$$\dot{H}_e = (\omega + d\omega) A \frac{P_{vap,s} - \phi_a P_{vap,a}}{R_{e,cl} + \frac{1}{f_{cl} h_e}} \tag{3.10}$$

In this equation, $\dot{H}_e$ is the enthalpy transfer caused by sweat evaporated from the skin [W]; $\dot{N}_e$ is the number of moles of water that evaporate from the skin surface [kmol/s]; M_{vap} is the molecular mass of water [kg/kmol]; $h_{fg}(T_s)$ is the enthalpy of evaporation of water at skin surface temperature; h_e is the evaporative mass transfer coefficient [W/m^2.°C]; $P_{vap,s}$ is the partial pressure of water vapor at the skin surface [Pa], customarily assumed to be the saturation pressure; $P_{vap,a}$ is the partial saturation pressure of water vapor in the air [Pa]. The multiplier factor ω, called the wet surface fraction, equals 0.06 when there is only diffusion (sweat is not present). When there is a change in environmental temperature, the term $d\omega$ can vary between 0 and 0.94, when the skin is completely wet with sweat. Finally, the term ϕ_a is the relative humidity of the air; $R_{e,cl}$ is the resistance to evaporation imposed by the clothing [Pa.m^2/W].

The enthalpy variation caused by breathing ($\Delta \dot{H}_{res} = \dot{H}_{ex} - \dot{H}_a$) has two physical mechanisms: one is due to the increased humidity of the expired air concerning the inspired one ($\dot{m}_{res}(w_{ex}h_{vap,ex} - w_a h_{vap,a})$), and the other results from the temperature difference between the inspired and expired air ($\dot{m}_{res}c_{p,a}(T_{ex} - T_a)$). The mass flow rate of inspired and expired air is directly proportional to the metabolism and can be evaluated by Eq. 3.12 obtained from [14].

$$\Delta \dot{H}_{res} = \dot{m}_{res}c_{p,a}(T_{ex} - T_a) + \dot{m}_{res}(w_{ex}h_{vap,ex} - w_a h_{vap,a}) \tag{3.11}$$

Where, $\Delta \dot{H}_{res}$ is the enthalpy flow rate associated with breathing [W]; $\dot{m}_{res}$ pulmonary ventilation [kg/s]; w_{ex} the absolute humidity of expired air [kg$_v$/kg$_a$]; w_a is the absolute humidity of inspired ambient air; $c_{p,a}$ the specific heat at constant air pressure [J/(kg.K)]; T_{ex} the exhaled air temperature [°C]; T_a the environmental inspired air temperature oC; $h_{vap,ex}$ the specific enthalpy of the vapor at the exhaled air temperature [J/kg]; $h_{vap,a}$ the specific enthalpy of the vapor at the inspired air temperature [J/kg].

$$\dot{m}_{res} = 1,43310^{-6} A_d \dot{M} \tag{3.12}$$

3.2.2 Exergy Transfer to the Environment

For the application of exergy analysis to the human body, the conditions of the environment under study have an air temperature T_0, pressure P_0, and relative humidity ϕ_0. Equation 3.13 shows the exergy associated with radiation, where the reference environment for this exergy is the mean radiant temperature (T_{mr}) of the surroundings [3]. Usually, the hypothesis of $T_{mr} = T_0$ is used.

$$\dot{B}_r = \left(1 - \frac{T_{mr}}{T_s}\right) \dot{Q}_r \tag{3.13}$$

The exergy flow rate associated with the heat exchanged by convection with the surroundings is in Eq. 3.14. The reference temperature for the convective exergy transfer is the air temperature which usually is at the environmental temperature $(T_0 = T_a)$

$$\dot{B}_c = \left(1 - \frac{T_0}{T_s}\right)\dot{Q}_c \tag{3.14}$$

Suppose the reference environment is the environment in which the body is. In that case, the exergy flow rate associated with evaporation remains as indicated by Eq. 3.15, in which the exergy associated with evaporation is evident (first term). The exergy associated with the change in saturated vapor concentration at the skin surface to the ambient water vapor concentration shows already the chemical exergy related to the chemical potential of the water vapor to diffuse into its natural composition, referring to the second term of the equation.

$$\dot{B}_e = \dot{m}_e\left[h_{lv} - T_0 s_{lv} + R_{vap} T_0 ln\left(\frac{P_{vap,s}}{P_{vap,a}}\right)\right] \tag{3.15}$$

The exergy variation resulting from breathing is related to the inspired and expired airflow that crosses the border of the control volume, as indicated by Eq. 3.16. The index i refers to the gases present in the atmosphere, for example, O_2, N_2, CO_2 e H_2O.

$$\Delta\dot{B}_{res} = \left(\sum \dot{m}_s b_s - \sum \dot{m}_e b_e\right)_{res} = -\sum_i \left(\dot{B}_{i,ex} - \dot{B}_{i,a}\right) \tag{3.16}$$

The exergy of the gases in inspiration is the exergy of the air itself, which is equivalent to zero if the environment is the reference. Hence, the exergy associated with breathing, Eq. 3.17, is only for expired air. Here, the term $y_{i,ex}$ is the mass fraction of each element at expiration. We will use y for a mass fraction in its gaseous and aqueous solution and the same for x on a molar basis.

$$\Delta\dot{B}_{res} = \dot{m}_{res}\left[c_{p,a}\left(T_{ex} - T_0 - T_0 ln\left(\frac{T_{ex}}{T_0}\right)\right) + \sum_{i=1}^{n} y_{i,ex} T_0 R_i ln\left(\frac{P_{ex,i}}{P_{0,i}}\right)\right] \tag{3.17}$$

3.3 Energy and Exergy Metabolism

The authors [4, 18, 19] point out that the primary practices of measuring metabolism are: (i) direct calorimetry and studied since the articles of [10] and [20]; (ii) indirect calorimetry (measurements of oxygen consumption and carbon dioxide production); (iii) indirect calorimetry, in which there is a control between the balance of calories consumed (glucose, fats, and proteins) and changes measured in the individual's body composition (fats and muscles) during weeks. This is the most complex method.

Regarding indirect calorimetry–(ii)–it is the most straightforward and typical way to calculate metabolism without using more complex devices and controlling diet and activities for an extended period.

Before an indirect calorimetry analysis, one must adequately define the physiological variable named respiratory quotient (RQ). According to the [14] standard, the RQ of a person in baseline conditions equals 0.83 (average of different nutrients oxidation). Equation 3.18 defines it as the ratio of the volume of carbon dioxide ($\dot{v}_{CO_2}$) expired to the volume of oxygen consumed $\dot{v}_{O_2}$ over time. Mountcastle [18] and Diener [19] performed a detailed analysis of indirect calorimetry and the calculation of metabolism. To this end, they defined three substances used in the body as the primary energy sources: carbohydrates, lipids, and proteins. For the calculation of energy and exergy metabolisms, the reactions of glucose (Eq. 3.19), palmitic acid (Eq. 3.20), and an amino acid with average thermodynamic composition (Eq. 3.21) [1].

$$RQ = \frac{\dot{v}_{CO_2}}{\dot{v}_{O_2}} = \frac{\dot{n}_{CO_2}}{\dot{n}_{O_2}} = \frac{\dot{m}_{CO_2} M_{O_2}}{\dot{n}_{O_2} M_{CO_2}} = 0.723\frac{\dot{m}_{CO_2}}{\dot{m}_{O_2}} \tag{3.18}$$

Figure 3.3 indicates an effort test using one type of calorimetry to assess metabolism. It is named indirect calorimetry and is the most common application to evaluate metabolism since it is undersized, low maintenance, and only needs a mask with the correct reading and measuring system of oxygen consumption and carbon dioxide production.

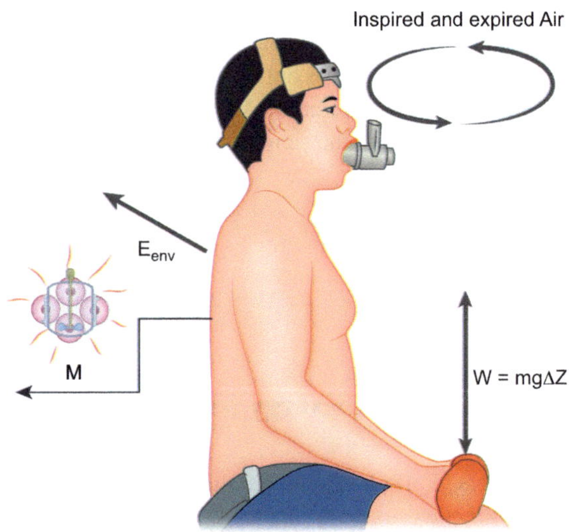

Fig. 3.3 A representative draw of an exercise test with respective measurements of oxygen consumption and carbon dioxide production. *Source* The author

Some authors [4] and [19] present oxidation reactions for amino acids. However, they use average values for the stoichiometric coefficients, a representative amino acid, though there is no transparency regarding the thermodynamic properties obtained. The authors [1] proposed equations for calculating metabolism (on an energy and exergy basis) based on indirect calorimetry measurements. Given that [21] assign the occurrence rate of amino acids in more than 1150 proteins found in nature, [1] obtained an amino acid with average composition ($C_{4.98}H_{9,8}N_{1.4}O_{2.5}$), from 95.8 % of the amino acids found in humans. Thus, the reaction 3.21 indicates the oxidation of this average amino acid until reaching the formation of carbon dioxide, liquid water, and urea. The authors excluded two because of the sulfur–S–in their composition and the difficulty of tracing their path in human metabolism. Recently, some authors considered the presence of this chemical compound; nevertheless, there are no details of the path of sulfur in the body [24].

$$C_6H_{12}O_6 + 6O_2 \rightarrow 6CO_2 + 6H_2O \tag{3.19}$$

$$C_{16}H_{32}O_2 + 23O_2 \rightarrow 16CO_2 + 16H_2O \tag{3.20}$$

$$C_{4.98}H_{9.8}N_{1.4}O_{2.5} + 5.135O_2 \rightarrow 4.28CO_2 + 3.5H_2O + 0.7CH_4N_2O \tag{3.21}$$

3.3.1 *Thermodynamic Properties and Equation of the Metabolism*

Based on the thermodynamic properties of the complete oxidation reactions of the various amino acids indicated in the literature, it was possible to assess the energy and exergy variation of the oxidation as they occur in the human body until the formation of urea. In this analysis, a substance of average properties reproduces amino acids. Carbohydrates and lipids are substances that may be represented by glucose and palmitic acid, respectively. The authors [25] calculated the chemical exergy of three fatty acids with 18, 20, and 22 carbons in the chain. The difference in chemical exergy per unit of mass is not more than 1%, whereas, on a molar basis, it will obviously have a difference. Of the three compounds explored in this work, the only thermodynamic property value that [23] does not supply is for amino acids, moreover, with the oxidation reaction that occurs inside the cell, producing urea. Applying the result obtained in [1], we can arrive at $\Delta h/\Delta b_{ch} = 1.04$. Thus, the change in chemical exergy for this author's data is determined to be 19777 kJ/kg.

Table 3.1 shows chemical exergy values (b_{ch}) and chemical exergy variation values (Δb_{ch}). These values are the input information to calculate the exergy variation of the molecules in the body. Nevertheless, the values are for the complete oxidation. Note that, as the temperature at which the oxidation reaction occurs is the same, the change in Gibbs free energy is equal to the change in exergy.

Tables 3.2 and 3.3 show the energy and exergy variations of the reaction of oxidation of nutrients as they occur in the human body (the oxidation is complete in the case of glucose and palmitic acid and partial for the amino acids). Based on the

Table 3.1 Chemical exergy of glucose, palmitic acid, and composition between chemical exergy of proteins and urea (kJ/kg). The properties are based on [23] and [26]

–	b_{ch} ou Δb_{ch} (kJ/kg)		
References	Glucose	Palmitic Ac.	Amino Ac.
Haynie [23]	16506	39141	19777
Aon et al. [26]	16516	39223	18317

Table 3.2 Enthalpy variation of glucose, palmitic acid, and amino acid during the reactions of oxidation (kJ/kg) as they occur in the body–until the formation of urea

–	Δh (kJ/kg)		
References	Glucose	Palmitic Ac.	Amino Ac.
Diener [19]	–15648	–39581	–18075
Haynie [23]	–15600	–39200	–19000
Aon et al. [26]	–15594	–39020	–

Table 3.3 Exergy variation of glucose, palmitic acid, and amino acid during the reactions of oxidation (kJ/kg) as they occur in the body–until the formation of urea

–	Δb (kJ/kg)		
References	Glucose	Palmitic Ac.	Amino Ac.
Haynie [23]	–15946	–38212	–18964
Aon et al. [26]	–15956	–38281	–17578

values of Δh and Δb (from the oxidation reactions as they occur in the body) and the procedure indicated by Mady and Oliveira [1], one can calculate the metabolism on an energy and exergy basis.

In this text, the equation based on studies by [1] will be used for exergy metabolism, as will be described in the next section. From the stoichiometric coefficients of the reactions of oxidation and their thermodynamic properties change, realizing energy and exergy, Mady and Oliveira-Junior [1] reached a system of Eqs. 3.22–3.24. We used the following convention: each gram of nitrogen excreted in the urine represents the oxidation of 6.25 g of amino acids, as in [26].

$$\dot{m}_{carb} = -2.14\dot{m}_{O_2} + 2.24\dot{m}_{CO_2} - 3.39\dot{m}_N \tag{3.22}$$

$$\dot{m}_{lip} = 1.14\dot{m}_{O_2} - 0.83\dot{m}_{CO_2} - 1.5\dot{m}_N \tag{3.23}$$

$$\dot{m}_{ami} = 6.25\dot{m}_N \tag{3.24}$$

Equations 3.25 and 3.26 define metabolism from the perspective of an energy and exergy basis. They are related to the consumption of nutrients and variations of thermodynamic properties during the oxidation of the nutrients. These reactions occur inside the human body. Hence they are a chemical potential change and part of energy dU_{body}/dt and exergy $d\mathbf{B}_{body}/dt$v variation of the human body (control volume)

$$-\dot{M} = \Delta\dot{H}_{oxi} = \dot{m}_{carb}\Delta h_{carb} + \dot{m}_{lip}\Delta h_{lip} + \dot{m}_{amin}\Delta h_{amin} \tag{3.25}$$

$$-\dot{B}_M = \Delta\dot{B}_{oxi} = \dot{m}_{carb}\Delta b_{carb} + \dot{m}_{lip}\Delta b_{lip} + \dot{m}_{amin}\Delta b_{amin} \tag{3.26}$$

For this text, the values of Eqs. 3.27 and 3.28 will be used given the discussion carried out in Mady and Oliveira-Junior [1] and the proximity of the results. Moreover, the amino acid with an average composition was part of the group results and significantly contributed to the literature. Note that only for this equation the values of $\dot{m}$ are in [g/s].

$$\dot{M} = 11.179\dot{m}_{O_2} + 2.502\dot{m}_{CO_2} - 1.580\dot{m}_N \tag{3.27}$$

$$\dot{B}_M = 9.558\dot{m}_{O_2} + 3.928\dot{m}_{CO_2} - 1.823\dot{m}_N \tag{3.28}$$

3.3.2 *Energy and Exergy Currency of the Body–ATP*

In this section, based on [1], there was no intention of calculating the chemical exergy of the molecules of ATP ($C_{10}H_{16}N_5O_{13}P_3$) and ADP ($C_{10}H_{15}N_5O_{10}P_2$) since these mix organic chains (adenosine) with inorganic elements (triphosphate)—this is of systems to define what would be the reference concerning the environment, the human body, or the external environment—here a tiny difference of strategy from engineering and biochemical thermodynamics. For the calculation of chemical exergy, Lems [27] started from the observation that the Gibbs energy of formation of the adenosine group is considered null under standard biochemical conditions and concentrated the calculations on the phosphate group. However, there was no mention of the binding energy of the phosphate group to the adenosine group, hence inorganic and organic bound.

As these compounds are either in the form shown on the left side of the Reaction 3.29 or on the right side–formation/hydrolysis. The phosphate group (P_i is, in fact, an ester of phosphoric acid (H_3PO_4, or after dissociation HPO_4^{-2}). This "mixture" is the point of the most significant dilemma in the exergy evaluation. The theory does not consider the bounds between organic and non-organic compounds. The analysis proposed in [28] is the one in which there is the best approach.

$$ADP + P_i \rightleftharpoons ATP + H_2O \tag{3.29}$$

Nevertheless, for simplicity's sake, the maximum available work possible to extract from this reaction (in this case, the hydrolysis of ATP) is the Gibbs free energy variation of these reactions. Therefore, it is enough to consider the variation as the exergy released inside the body without considering the chemical exergy of each chemical compound of the living beings between the natural conditions of the body and the reference environment, given that exergy is the measure of this deviation. The human body is in constant disequilibrium (thermal, mechanical, and chemical) with the environment. Only the variation may receive special attention since there is a deviation between biological and natural conditions inside the human body concerning the actual environment; in this text, we will treat it as $\dot{W}_{MAX}$.

The only adaptation is to consider the deviation properly as it occurs inside the human body and the difference in concentrations with the reference environment, which differs from the biological reference. Considering that for the exergy calculation, the temperature and pressure conditions of the biochemical compounds are the same as the reference environment (that is, the physical exergy in the oxidation reactions is neglected), the variation of the Gibbs free energy of the reactions is the variation of exergy itself. More details about this thermodynamic/mathematical equivalence are in the [32]. To treat metabolism at the cellular scale, one must consider the hydrolysis or formation of the ATP molecule, shown in Reaction 3.29. This reaction is the energy source for all body activities. Hence, the ATP is colloquially named in the biochemical literature "energy coin". The authors [21, 27, 29] show that the standard biochemical and actual cell conditions are not always the same. As described in [1], it can be seen that the exergy variation under natural conditions (–56 kJ/mol) is almost double that of biochemical conditions (–30.5 kJ/mol) for ATP hydrolysis.

To consider the effect of acid dissociation, one must find the equilibrium constants k_i. Lems [27] found the values of pk_i ($pk_i = -log(k_i)$) for the seven constants. The author recalculated the equilibrium constants to obtain the effect of ionic interactions. However, this text will use the results proposed by [30, 31]. The authors calculated the Gibbs energy of formation and the enthalpy of formation for several biochemical compounds under conditions "close to physiological conditions", which are: T = 298.15 K, P = 1 atm, pH = 7, pMg = 3, and 0.25 mol/L ionic strength. Thus, for these conditions, they obtained that the change in Gibbs free energy for ATP hydrolysis is equivalent to –32.48 kJ/mol, with only the correction for the concentrations of reactants and products remaining. In this case, as well as for [27], the effects mentioned above are small compared to the absolute value of the Gibbs energy change in the ATP hydrolysis reaction (approximately 1 to 2 kJ/mol). In the case of formation properties, it becomes relevant. Nevertheless, we are dealing with energy and exergy variations.

Finally, the concentrations of ATP, ADP, and P_i from the biochemical reference must be changed to the natural state of the cell, as indicated in Eq. 3.30 for ATP hydrolysis, which is equivalent to 1, 85, 0.14 and 1.0 mmol/L, respectively. These values are available in [27] and [29] for red blood cells (conditions close to the hyaloplasm of human cells) and were obtained by [33]. Although [21] shows an ADP concentration greater than 0.25 mmol/L, it states that this value can be much lower. Therefore, the change in Gibbs free energy becomes –56.1 kJ/mol (the effect of concentrations equals –23.51 kJ/mol, being the most relevant adjustment).

$$\Delta\bar{g} = \Delta\bar{g}'^0 + \bar{R}T_0 ln\left(\frac{[ADP][P_i]}{[ATP]}\right) = -56,1\,\text{kJ/mol} \tag{3.30}$$

Relying on the concept of cellular metabolic efficiency (η_M) [34] indicated in Eq. 3.31, which is the ratio between the exergy released in the hydrolysis of ATP and in the complete combustion of nutrients.

$$\eta_M = \frac{\Delta\dot{G}_{ATP}}{\Delta\dot{G}_{oxi,nutrients}} \tag{3.31}$$

One can write the maximum work the body can do from the energy contained in ATP molecules, as in Eq. 3.32. Note that this can be assessed by evaluating the number of moles of ATP that undergo hydrolysis ($\dot{n}_{ATP}$) or the consumption of nutrients (glucose, lipids, and carbohydrates).

$$\dot{W}_{max} = \dot{n}_{ATP}\left|\Delta\bar{g}_{ATP}\right| = \sum_{i=1}^{3}\eta_{M,i}\left|\Delta\dot{G}_i\right| \tag{3.32}$$

Where $\dot{W}_{max}$ is the maximum work available from the chemical energy contained in ATP molecules, $\dot{n}_{ATP}$ is the number of mols of ATP that underwent hydrolysis, Δg_{ATP} is the change in Gibbs free energy in the hydrolysis of ATP, ΔG_i is the change in Gibbs free energy of the oxidation reaction of the nutrient i (i refers to glucose, lipids, and amino acids) and $\eta_{M,i}$ is the metabolic efficiency of the oxidation reaction.

3.3.2.1 Results of Energy and Exergy Metabolism

Based on methods of Sect. 3.3.1 the number of ATP that are formed for each nutrient, it is possible to calculate the metabolic efficiency. This value is 62% for glucose, 61% for palmitic acid, and 20% for amino acids. Figure 3.4 shows a comparison of the metabolism on an energy and an exergy basis, maximum available work, and exergy released to the body associated with metabolism. These results were obtained by [1] for one runner on a treadmill, indicating that the body maintains approximately 60% of the exergy content of the nutrients in the chemical bond of the phosphate group in ATP, and this exergy is available to perform work. If the metabolism was considered as a heat source, this efficiency would be only 5%, and the work produced by the subject would be higher than $\dot{B}_{Q_M}$, showing a clear violation of the second law of thermodynamics.

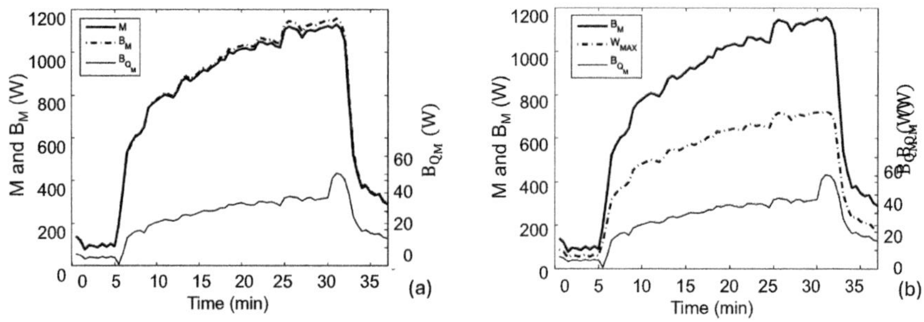

Fig. 3.4 Exergy conversion process in cellular metabolism. From nutrient oxidation, ATP formation and the use of ATP. Based on the figures proposed in [35]. **a** Bicycle test; **b** weight lifting

3.3.3 Anaerobic Metabolism

In order to understand the quality of energy conversion during exercises [8] and even pathologies [36], it is necessary to evaluate the behavior of muscles' energy metabolism. One of the critical factors to be analyzed is metabolic efficiency (Eq. 3.31, where the metabolism can be aerobic or anaerobic). The chemical exergy of glucose is 2971.1 kJ/mol (Table 3.1). However, to complete the calculation, it is necessary to calculate the chemical exergy of lactic acid. Bearing in mind that $\Delta\dot{G}_{reac} = \Delta\dot{B}_{reac}$, as evidenced by [1]).

Given the chemical exergy obtained by [8] for lactic acid of 1339.6 kJ/mol. The change in the ability to perform any kind of work (or change in the Gibbs free energy of the anaerobic reaction) is equivalent to 226.44 kJ/mol. The aerobic and anaerobic glucose oxidation reactions in the human body are indicated in Table 3.4, as well as the amount of ATP formed for each mol (or kmol) of glucose oxidation and their Gibbs free energy variation values. It is worth mentioning that glucose was chosen for the analysis since, among the three types of compounds present in the human diet, glucose, lipids, and amino acids, this is the only one that undergoes anaerobic oxidation to provide fast energy when the presence of oxygen is scarce; however, with a thermodynamic cost of low efficiency.

The anaerobic oxidation forms lactic acid, or lactate, whose molecular structure presents hydrogen and carbon bonds, being a compound with high chemical exergy, $\bar{b}_{ch} = 1339.6$ kJ/mol. Eventually, the respiratory coefficient (RQ) is an indicator that

Table 3.4 Aerobic and anaerobic reactions of glucose oxidation with respective values of Gibbs free energy change. Obtained in [8]

Reaction	Equation	ATP	$-\Delta\bar{g}_f^o$ [kJ/mol]
Aerobic	$C_6H_{12}O_6 + 6\,O_2 \rightarrow 6\,CO_2 + 6\,H_2O$	32	2,872.0
Anaerobic	$C_6H_{12}O_6 \rightarrow 2\,C_3H_6O_3$	2	226.44

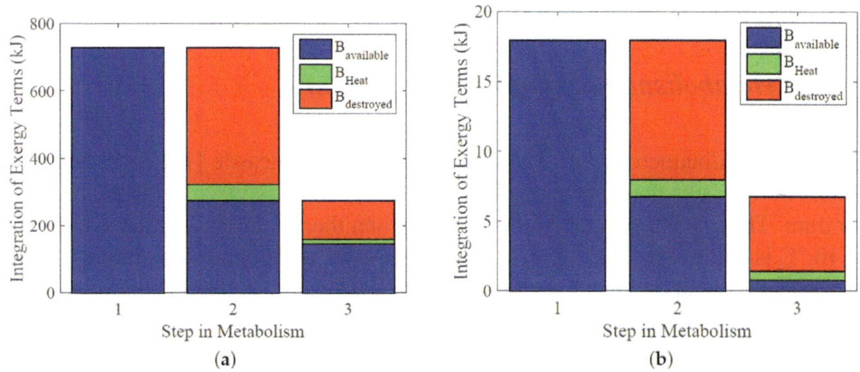

Fig. 3.5 Exergy conversion process in cellular metabolism. From nutrient oxidation, ATP formation and the use of ATP. Based on the figures proposed in [8]. **a** Bicycle test; **b** weight lifting

there is a greater consumption of nutrients via anaerobic metabolism if it is greater than 1; to determine the amount of lactate formed, the stoichiometric relationship $\dot{n}_{lactate} = 2 \times \dot{n}_{glucose}$, it also shows the low degradation of the glucose. The efficiency of this reaction is lower than the complete oxidation; however, as it only divides the glucose, with a cost of 2 ATP, it must be faster to meet demand in the absence of oxygen; [38] conducted the same reasoning, although referring to a cell with pathologies. The overall production is 4 ATP's with a net positive balance of 2 mols of ATP.

Figure 3.5 indicates the difference in metabolic energy conversion and their respective losses in each process for aerobic [35] and anaerobic (Fig. 3.3) [8] exercise tests. For Fig. 3.5a, one bicycle of the test conducted by one of the cyclists, the total energy consumption in metabolic rate was 730 kJ (around 175 kcal) in an exhaustion test, whereas for the weight lifting the energy expenditure was lower, 18 kJ, due to the nature and duration of the exercise protocol, as in Fig. 3.5b. The main conclusion is that the percentage of $\dot{W}_{MAX}$ that becomes real performed power $\dot{W}$, being the aerobic exercise more efficient in using the energy content of nutrients. The results of [8] also indicate that, by applying Eq. 3.18, the respiratory quotient of Fig. 3.5a remains around 1.0 (predominantly aerobic), whereas in Fig. 3.5, it stays between $1.0 \leq RQ \leq 1.2$. Eventually, the question regarding human health and the consequences of both types of exercise in exergy literature is still open.

Figure 3.5b depicts the movement that represents a series of weight lifting (biceps curl). This experimental protocol aimed to propose the most suitable exercise for applying the exergy analysis; nevertheless, some challenges may arise in defining the performed work by the muscle, as the complete movement would result in a liquid work equivalent to zero. Therefore, we only considered the energy content to lift the weight. It is worth noting that the remaining details of this figure will be discussed in future sections. The most significant result is the low exergy available (work performed) compared to the exergy of metabolism.

3.4 Thermoregulatory System of the Human Body

3.4.1 Metabolism and Internal Temperature

The authors Albuquerque and Yanagihara [7] and Mountcastle [18] described that metabolic reactions have their speed altered when there is variation in tissue temperature. The ratio of two metabolism rates when there is a temperature difference of 10 °C is called the "Q_{10} effect". This increase in metabolic rate is a function of average tissue temperature (T_t), average tissue temperature at basal conditions ($T_{t,0}$) increasing the basal metabolism (M_0), demonstrated in Eq. 3.33, also represented by Fig. 3.6.

$$\frac{\dot{M}_{Q10}}{M_0} = 2^{(T_t - T_{t,0})/10} \tag{3.33}$$

3.4.2 Control System

In their studies, Ferreira and Yanagihara [9] considered temperature signals from the hypothalamus (in the brain model) and the skin. The model employed has its bases in experimental research (determination of constants and equations). The control systems are vasomotor mechanisms, sweating, and chills. Mady et al. [39] obtained that the constants of the thermoregulatory system that present conditions close to the minimum exergy destroyed rate and maximum exergy efficiency are those obtained

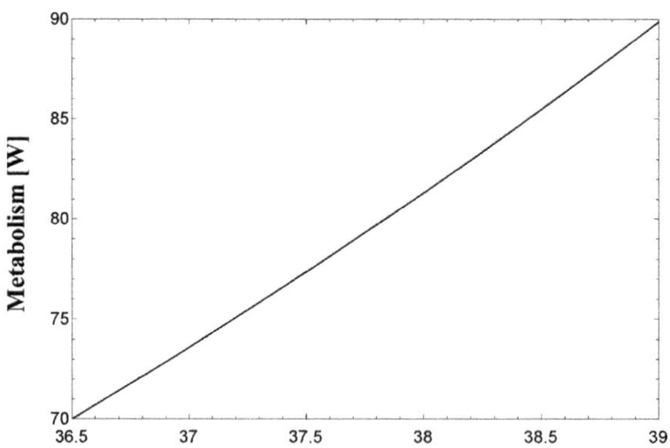

Fig. 3.6 Variation of metabolism as a function of temperature according to Eq. 3.33. *Source* Own author

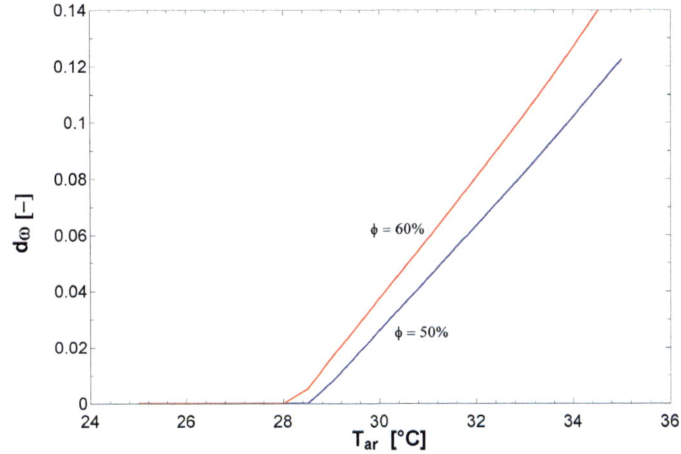

Fig. 3.7 Skin wetness ω as a function of air temperature to a model of the thermal system of the human body proposed by [12]

experimentally from the literature; that is, the constants that correctly represent the behavior of the control system of a human being.

Svage et al. [40] proposed Eq. 3.34, whose coefficient calculus is from simulations that respond similarly to the observed behavior of the human body for a provided anatomy.

$$\Delta\dot{\omega}_{bl} = k_1(T_{hy} - T_{hy,0}) + k_2(\overline{T}_s - \overline{T}_{s,0}) \tag{3.34}$$

Where, $\Delta\dot{\omega}_s$ is the rate of blood perfusion to the skin ($814.10^{-6} \leq \Delta\omega_s \leq 3345$ $m^3/(m^3.s)$); T_{hy} is the hypothalamus temperature at an instant t; $T_{hy,0}$ is the hypothalamus temperature at thermal neutrality; $\overline{T}_s$ is the average skin temperature at an instant t; $\overline{T}_{s,0}$ is the reference temperature at thermal neutrality; k_1 and k_2 are the proportional control system constants, equal to 1810.10^{-6} $m^3/(m^3.s)$, 181.10^{-6} $m^3/(m^3.s)$, respectively.

Nadel et al. [41] propose an equation for the sweating control mechanism according to Eq. 3.35. In this equation, the effect of local changes in the skin, e.g., sweating, and the average effect of the whole body are considered.

$$\dot{H}_{e,sw} = A.\left[k_3(T_{hy} - T_{hy,0}) + k_4(\overline{T}_s - \overline{T}_{s,0})\right] exp\left(\frac{\overline{T}_{s,i} - \overline{T}_{s,0}}{10}\right) \tag{3.35}$$

Where, $\dot{H}_{e,sw}$ is the sweat evaporation rate from the skin [W]; $\overline{T}_{s,i}$ is the average skin surface temperature in element i. The constants k_3 and k_4 are 197 $W/(m^2.K)$; and 23 $W/(m^2.K)$, respectively.

When applying Eq. 3.35 together with the definition of skin wetness, it is possible to obtain Fig. 3.7. It is indicated ω as a function of air temperature to a model of the thermal system of the human body proposed by [12]. This figure demonstrates the skin wetness as a function of relative humidity for the simplified model of the thermal system proposed in Mady et al. [3] for two relative humidities. Note that if the system has a value of $\omega = 0.2$, it means that 20% of the surface is covered by sweat. This factor will directly influence the water that evaporates at the skin, which, together with diffusion through the skin, provides the enthalpy of vaporization due to sweating.

It is possible to assess the increase in the metabolism due to shivering mechanisms according to Eq. 3.36, elaborated by [42]. Again, the skin and hypothalamus (representative of the internal) temperatures are considered.

$$\Delta \dot{M}_{sh} = k_5(T_{ty} - T_{ty,0}) + K_6(\overline{T}_s - \overline{T}_{s,0}) + k_7 \Delta Q. \tag{3.36}$$

Where, $\Delta \dot{M}_{sh}$ is the rate of change of metabolism caused by chills (W); k_5, k_6 and k_7 are constants of the thermoregulatory system, equal to 250 W/K, 40 W/K, and 0.06, respectively. Again, these constants are validated later to be the ones that cause minimal destroyed exergy and maximum exergy efficiency as discussed by Mady et al. [39].

References

1. Mady CEK, Oliveira-Junior S (2013) Human body exergy metabolism. Int J Thermodyn 16(2):73–80
2. Mady CEK, Albuquerque C, Fernandes TL, Hernandez AJ, Saldiva PHN, Yanagihara JI, Oliveira-Junior S (2013) Exergy performance of human body under physical activities. Energy 62:370–378
3. Mady CEK, Albuquerque C, Fernandes TL, Hernandez AJ, Yanagihara JI, Oliveira-Junior S (2014) Human body exergy analysis and the assessment of thermal comfort conditions. Int J Heat Mass Transfer 77:577–584
4. Batato M, Deriaz O, Borel L, Jequier E. Analyse exergétique, théorique et expérimentale, du corps human. Entropie 26:120–130
5. Cenzi JR, Albuquerque C, Mady CEK (2018) The effect of carbon monoxide on the exergy behavior of the lungs. Bioengineering 5(4):108
6. Cenzi JR, Albuquerque C, Mady CEK (2019) Phenomenological and thermodynamic model of gas exchanges in the placenta during pregnancy: a case study of intoxication of carbon monoxide. Int J Environ Res Publ Health 16(21):4138
7. Albuquerque C, Yanagihara JI (2020) An integrated model of the thermoregulatory and respiratory systems of the human body. Int Commun Heat Mass Transfer 116:104683
8. Spanghero GM, Albuquerque C, Lazzaretti FT, Hernandez AJ, Mady CEK (2018) Exergy analysis of the musculoskeletal system efficiency during aerobic and anaerobic activities. Entropy 20(2):119
9. Ferreira MS, Yanagihara JI (2009) A transient three-dimensional heat transfer model of the human body. Int Commun Heat Mass Transfer 36(7):718–724
10. Hardy JD, Du Bois EF, Soderstrom GF (1938) Basal metabolism, radiation, convection and vaporization at temperatures of 22 to 35 C.: Six figures. J Nutrition 15(5):477–497
11. Martinez-Garcia M, Une RY, Oliveira-Junior S, Mady CEK (2018) Exergy Analysis and Human Body Thermal Comfort Conditions: Evaluation of Different Body Compositions. Entropy 20:265

12. Molliet DS, Mady CEK (2021) Exergy analysis of the human body to assess thermal comfort conditions: Comparison of the thermal responses of males and females. Case Stud Thermal Eng 25:100972
13. Molliet DS, Mady CEK (2024) Investigation of a Building with Male and Female Residents to Achieve Equality of Thermal Sensation Associated with Rational Use of Exergy and Energy. Buildings 14(4):1149
14. American Society of Heating, Refrigerating, Air-Conditioning Engineers (2005) ASHRAE handbook
15. ISO7730 (2005) Ergonomics of the thermal environment-Analytical determination and interpretation of thermal comfort using calculation of the PMV and PPD indices and local thermal comfort criteria
16. DuBois D (1916) A formula to estimate the approximate surface area if height and weight be known. Arch Intern Med 17:863–871
17. Dear RJ, Arens E, Hui Z, Oguro M (1997) Convective and radiative heat transfer coefficients for individual human body segments. Int J Biomet 40:141–156
18. Mountcastle VB (1974) Medical physiology. CV Mosby
19. Diener JRC (1997) Calorimetria indireta. Revista da Associação Médica Brasileira 43:245–253
20. Hardy JD, Milhorat AT, Du Bois EF, Soderstrom GF (1941) Basal Metabolism and Heat Loss of Young Women at Temperatures from 22 °C to 35 °C: Clinical Calorimetry No. 54; Five Figures. J Nutrition 21(4):383–404
21. Nelson DL, Lehninger AL, Cox MM (2008) Lehninger principles of biochemistry. Macmillan
22. Igarashi TL, Fernandes TL, Hernandez AJ, Mady CEK, Albuquerque C (2022) Behavior of skin temperature during incremental cycling and running indoor exercises. Heliyon 8:10
23. Haynie DT (2001) Biological thermodynamics. Cambridge University Press
24. Yildiz C, Köhler K, Wasserfurth P, Özilgen M (2024) Assessment of entropy accumulation in human subjects when exposed to low energy availability. Heliyon 10(17)
25. Sorguven E, Özilgen M (2010) Thermodynamic assessment of algal biodiesel utilization. Renew Energy 35(9):1956–1966
26. Aon MA, Cortassa SDC, Iglesias AA, Lloyd D (2002) An Introduction To Metabolic And Cellular Engineering. World Scientific
27. Lems S, (2009) Thermodynamic explorations into sustainable energy conversion. Learning from living systems. PhD thesis, Technische Universiteit Delft
28. Valero VC, Valero AV (2014) Thanatia: The Destiny of The Earth's Mineral Resources-A Thermodynamic Cradle-to-cradle Assessment. World Scientific
29. Hammes GG (2000) Thermodynamics and kinetics for the biological sciences. John Wiley & Sons
30. Alberty RA, Goldberg RN (1992) Standard thermodynamic formation properties for the adenosine 5'-triphosphate series. Biochemistry 31(43):10610–10615
31. Alberty RA (1998) Calculation of standard transformed Gibbs energies and standard transformed enthalpies of biochemical reactants. Arch Biochem Biophys 353(1):116–130
32. De Nevers N, Seader JD (1980) Lost work: A measure of thermodynamic efficiency. Energy 5:757–769
33. Minakami S, Yoshikawa H (1965) Thermodynamic considerations on erythrocyte glycolysis. Biochem Biophys Res Commun 18(3):345–349
34. Annamalai K, Silva C (2012) Entropy stress and scaling of vital organs over life span based on allometric laws. Entropy 14(12):2550–2577
35. Mady CEK, Igarashi TL, Albuquerque C, Santos-Silva PR, Fernandes TL, Hernandez AJ (2019) Exergy efficiency on incremental stationary bicycle test: A new indicator of exercise performance? J Brazilian Soc Mech Sci Eng 41:1–11
36. Henriques IB, Oliveira-Junior S (2020) Tumour cells and life expectancy: an exergy approach. Int J Exergy 31(2):103–119
37. Szargut J, Morris DR, Steward FR (1988) Exergy analysis of thermal, chemical, and metallurgical processes. Hemisphere Publishing Corporation, New York

38. Luo LF (2009) Entropy production in a cell and reversal of entropy flow as an anticancer therapy. Front Phys China 4:122–136
39. Mady CEK, Ferreira MS, Yanagihara JI, Saldiva P, Oliveira-Junior S (2011) Second law of thermodynamics and human body. Therm Eng 10(1–2):88–95
40. Savage MV, Brengelmann GL (1996) Control of skin blood flow in the neutral zone of human body temperature regulation. J Appl Physiol 80(4):1249–1257
41. Nadel ER, Bullard RW, Stolwijk JA (1971) Importance of skin temperature in the regulation of sweating. J Appl Physiol 31(1):80–87
42. Gordon RG, Roemer RB, Horvath SM (1976) A mathematical model of the human temperature regulatory system-transient cold exposure response. IEEE Trans Biomed Eng 6:434–444
43. Asghari M, Brouwer J (2019) Integration of a solid oxide fuel cell with an organic rankine cycle and absorption chiller for dynamic generation of power and cooling for a residential application. Fuel Cells 19(4):361–373
44. Herold KE, Radermacher R, Klein SA (2016) Absorption chillers and heat pumps. CRC Press

Chapter 4
Experiments of Athletes' Performance During Physical Exercises

4.1 Experiments

All tests were conducted following approved protocols at the Faculty of Medicine of the University of São Paulo, with ethical clearance from the Ethics Committee for Analysis of Research Projects (CAPPesq) under registration numbers 16507 and 2477702. It is worth noting that such protocols are designed to induce exhaustion in participants.

As discussed by Mady et al. [1], the maximum oxygen consumption, denoted as $\dot{v}_{O_{2,max}}$, is defined by Mountcastle [2] as the highest rate of energy released by metabolism to perform work using oxygen. For patients to reach maximum oxygen consumption, specific protocols need to be employed. Figure 4.1 illustrates the procedure for calculating this physiological variable for runners with varying levels of functional capacity. It usually increases every 2 or 3 min during strenuous activity. There comes a point where the demand for work exceeds the body's ability to oxidize nutrients, leading to an increased rate of anaerobic oxidation. The intensity of physical activity increases until reaching a point where oxygen consumption does not change, regardless of the speed; this point is the $\dot{v}_{O_{2,max}}$.

Figure 4.2 illustrates the experimental procedure for evaluating runners during an exhaustion test. Initially, the protocol was adjusted individually for each participant, e.g., initial speed. Such initial results served as the basis for proposing a new protocol; hence, the test evolved for every participant to have the same starting speed. Another weakness of the treadmill is the calculation of the power exerted by the runner, according to the thermodynamic definition of "having the same effect as lifting a weight into the environment." The subsequent phase involved weightlifting, as depicted in Fig. 3.3, as emphasized by [5], but such tests were significant for determining the point at which metabolism becomes anaerobic. Finally, as highlighted in [3], the bicycle ergometric test offers the advantage of measuring the performed power, which becomes a known variable. Furthermore, it satisfies the thermodynamic definition of work, as illustrated in Fig. 4.3.

C. E. Keutenedjian Mady, *Exergy Analysis*, Green Energy and Technology,
https://doi.org/10.1007/978-3-032-01646-1_4

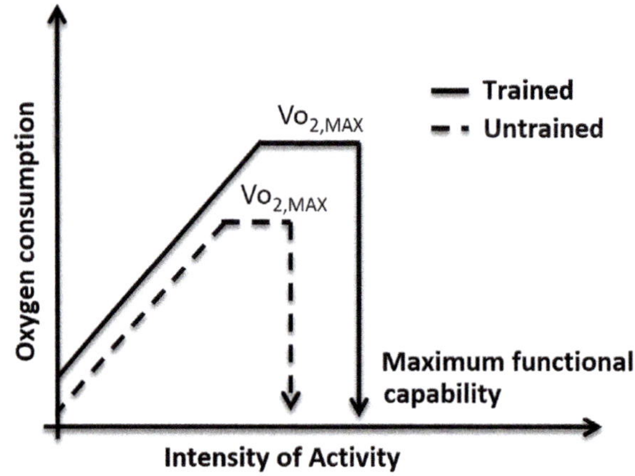

Fig. 4.1 Representation of the oxygen consumption over time and intensity of physical exercise for calculating the maximum oxygen consumption. *Source* Own author, adapter from literature, [2]

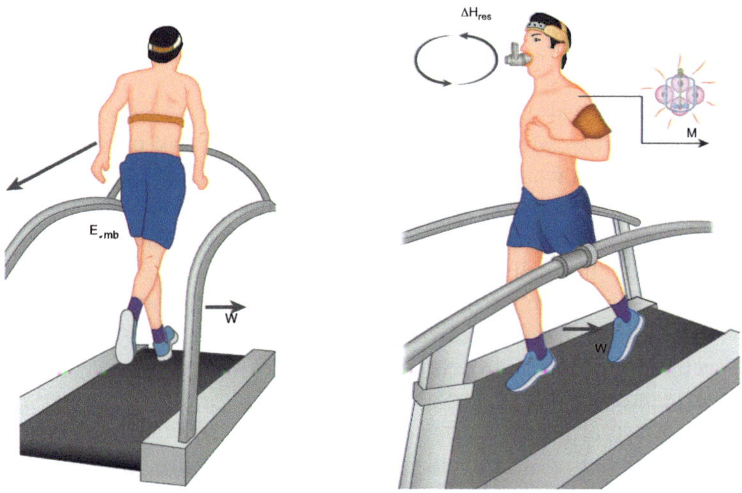

Fig. 4.2 Graphical representation of ergospirometry activity on treadmill. *Source* Own author, also based on [3]

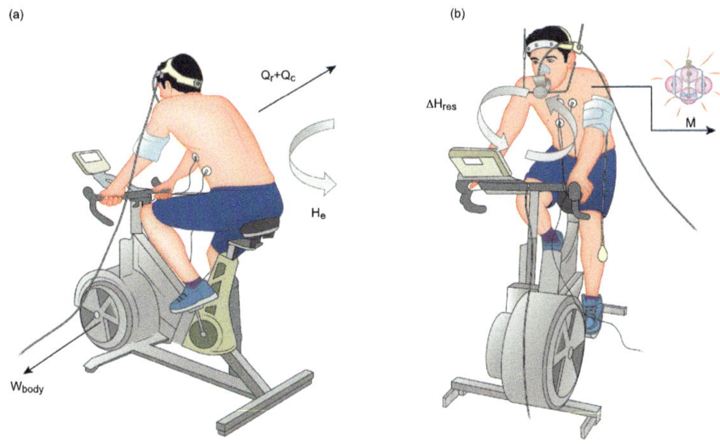

Fig. 4.3 Graphical representation of ergospirometry activity on treadmill. *Source* Own author, also based on [3]

4.2 Phenomenological Behavior of the Human Body Under Physical Activities

Depending on the activity the body engages in, there is a potential for an increase in the power generated through energy (or exergy) metabolism. In the subsequent sections, a comparison of three activities—running, cycling, and weightlifting—will be conducted. The human body adapts differently to each activity, sweating, vasodilation/vasoconstriction, with unique responses. Therefore, aspects such as temperature regulation, energy transfer to the environment, and other physiological factors will be discussed for each activity. Three types of tests were conducted in different time periods:

- **Test 1**: Eleven male subjects participated in this study, all residing in the city of São Paulo or its surrounding areas. Their average age was 31.4 $\pm$ 10.1 (mean $\pm$ standard deviation) years, with an average mass of 73.3 $\pm$ 7.8 kg; height of 1.76 $\pm$ 0.07 m, and body surface area of 1.89 $\pm$ 0.11 m^2. All subjects were runners with varying levels of training, with seven of them training four or more times a week, and four training less than four times a week. During the experiments, the runners wore only shorts and running shoes. The assessment began with a 5-min control period while the subject stood on the treadmill. Then the subject ran for 3 min to warm up with 30% of their long-distance training speed. This specific point posed a challenge in the analyses, since the protocol varied based on the individual's physical capacity, and some uniformity was necessary to ensure consistency in the test behavior. After that, the speed was set to 70% of the training speed and was incremented by 1 km/h every 4 min until the runner reached exhaustion.

Table 4.1 Subjects characteristics: subject ID, birth sex, mass, lean mass, height, and maximum oxygen consumption for running and cycling. The tests were carried out at the Sports Medicine Group of the Department of Orthopedics and Traumatology, University of São Paulo Medical School, São Paulo, Brazil. As a result, the Research Ethics Committee approved the project of the Clinics Hospital of the University of São Paulo Medical School, São Paulo, Brazil (CAAE 2.477.702)

ID	Sex	Age	m	m_{lean}	h	$\dot{v}_{O_2,max,running}$	$\dot{v}_{O_2,max,cycling}$
–	–	[Years]	[kg]	[kg]	[m]	[ml min^{-1} kg^{-1}]	[ml min^{-1} kg^{-1}]
1	M	41	73.2	34.2	1.78	51.0	56.6
2	M	47	80.9	37.5	1.80	43.5	42.0
3	M	34	67.2	32.0	1.70	64.1	53.3
4	M	41	56.8	27.2	1.60	38.9	55.4
5	F	30	57.5	23.2	1.60	46.8	44.8
6	F	47	56.3	20.4	1.55	50.1	41.9
7	F	27	64.3	27.0	1.57	41.9	43.2
8	F	30	66.5	28.8	1.65	44.8	38.5
Average		37	65.3	28.8	1.66	47.6	47.0
Standard deviation		8	8.7	5.6	0.09	7.8	7.0

Throughout the experiment, the treadmill was inclined by 1% to simulate the energy expenditure equivalent to outdoor running [1].

- **Test 2**: The tests were conducted between February and May 2018; while the hot climate in São Paulo would not interfere as much as in extreme conditions, the laboratory had climate control infrastructure [1, 3, 6]. Nonetheless, there were variations in environmental temperature due to the size of the environment. For the running tests, the average air temperature was 24.2 ± 3.8 °C, with a relative humidity of $63 \pm 8\%$, and 26.6 ± 3.1 °C with a relative humidity of $58 \pm 5\%$ for cycling tests. No significant differences were found between the environmental conditions during the cycling and running tests (unpaired t-test, $p = 0.21$ for temperature and $p = 0.13$ for humidity). The barometric pressure at the site was 688 mmHg. Eight amateur athletes, who regularly trained and participated in sports competitions, were included in the study. Each participant performed running and cycling exercises on different dates. The anthropometric characteristics of the individuals, as well as their maximum oxygen consumption, are summarized in Table 4.1. All athletes underwent the same warm-up and recovery regimen, although the duration of the exercise varied for each individual. The recovery phase began as soon as the athlete indicated exhaustion, based on their subjective perception of exertion using the Borg Scale [7].
- **Test 3**: Weight lifting (biceps curl): a continuous series of lifting (76 repetitions) was performed using one arm (the forearm is considered to elevate the mass to 0.26 m) and a mass of 4 kg until the participant reached exhaustion. This experimental protocol was designed to determine the most suitable exercise for applying the exergy analysis [5].

4.2.1 Work Performed by the Body

In order to accurately apply the first law of thermodynamics to the human body, it is essential to obtain certain parameters from the ergospirometry test. The most debated term refers to the work performed (or power performed), as emphasized by [1, 3, 6]. Therefore, it is crucial to evaluate the effects of the Work Calculation Equation on the thermodynamic behavior of the human body during treadmill exercise, weightlifting, and bicycle ergometer exercise.

In the following sections, a logical analysis was conducted to effectively demonstrate the research field formation and consolidation. In the biomedical literature, there is a discrepancy regarding the calculation of the First Law efficiency of the human body calculated by [8] decreases with increasing speed, from approximately 60 to 15% between 3.9 and 9.4 m/s. However, [9] demonstrated that the efficiency remains relatively constant (55% $\pm$ 12.7). In contrast, [10] determined an efficiency of 44% for runners at a speed of 3.6 m/s; and [11] found an efficiency of 44% for running on an inclined treadmill at a speed of 2.2 m/s. Such variation of data can be attributed to differences in the calculation of the metabolism when in anaerobic conditions [8] and the method used to calculate the power, ranging from 343 to 1650 W for speeds between 3.6 and 3.9 m/s, as reported by ref [12].

In the medical literature, different methods for assessing power on a treadmill were found, as outlined by Eqs. 4.1 to 4.2. It is also important to note that these equations take into account a 1% incline of the treadmill. The sources for such equations include [12–14]. Equation 4.1 represents the maximum power that the body could obtain from the ATP molecule, while Eqs. 4.3 and 4.4 indicate the work per unit of internal time and the outermost inner sum proposed by [12]. Finally, [13] proposes a model of the body as a material point ascending at a constant rate, as shown in Eq. 4.2. In these Equations, V represents the speed of the runner in m/s, and m_{body} represents the mass of the body in kg. The term $\Delta \dot{G}_{ATP}$, as discussed previously, refers to the Gibbs free energy release during ATP hydrolysis.

$$\dot{W} = \dot{W}_{MAX} = -\Delta \dot{G}_{ATP} \tag{4.1}$$

$$\dot{W} = \dot{W}_{Ponto,Mat} = m \times g \times V \times sen(\theta) \tag{4.2}$$

$$\dot{W} = \dot{W}_{Int} = \frac{4,184}{60} \times m_{corpo} \times (9,42 + 4,73V) \tag{4.3}$$

$$\dot{W} = \dot{W}_{Int+Ext} = \frac{4,184}{60} \times m_{corpo} \times (9,42 + 4,73V + 0.266V^{1.993}) \tag{4.4}$$

One of the subjects, who achieved a maximum velocity of 15 m/s, had the performed power calculated using Eqs. 4.1 to 4.2. The results are shown in Fig. 4.4. It is evident that the power calculated using Eq. 4.1 yielded smaller results. A comparison between Eqs. 4.1 and 4.2 indicates that the method proposed by [15] provides closer results to the maximum work performed obtained by ATP hydrolysis. The

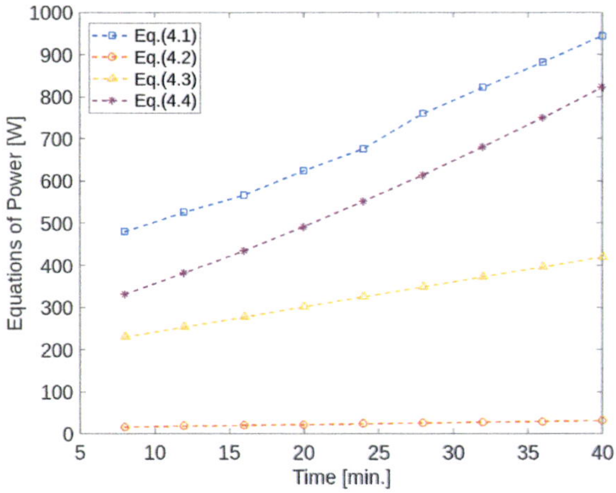

Fig. 4.4 Performed power as a function of time calculated using Eqs. 4.1–4.2

only difference is that the first equation is based on indirect calorimetry results, and the second is based on velocity and biomechanics calculation, resulting in different magnitudes.

The choice of several cyclists is related to the previous knowledge of the performed work. The collected data allowed us to evaluate the metabolic exergy, the maximum available work in ATP molecules, and the actual performed power. One conclusion drawn from the comparison in Fig. 4.5 is that the efficiency of aerobic exercise is higher than weight lifting. However, a careful interpretation of this result reveals that aerobic activities involve a larger muscle volume, leading to increased metabolic exergy and maximum available work for performing the required power. Nevertheless, this section aims to present three types of results, and the only one directly determined without any assumptions is the bicycle exercise test. This is relevant for future experiments, as shown in Fig. 4.5. Moreover, each energy conversion process decreases the quality of the energy, decreasing the exergy. Hence, for the body to obtain ATP, it is necessary to destroy energy. Nevertheless, this molecule is available through the body cells for fast use and production.

4.2.2 Energy and Exergy Metabolism

In **Test 1**, the oxygen consumption ($\dot{m}_{O_2}$) and carbon dioxide production ($\dot{m}_{CO_2}$) were measured. By using Eq. 3.18, it is possible to convert the respiratory (RQ) coefficient from a molar basis to a mass basis. Figure 4.6 presents the results for the individual who achieved a speed of 15 m/s (shown on the secondary ordinate axis), where it is possible to observe that, unlike basal conditions where the respiratory coefficient is usually around 0.83, during physical exercises it increases and approaches unity.

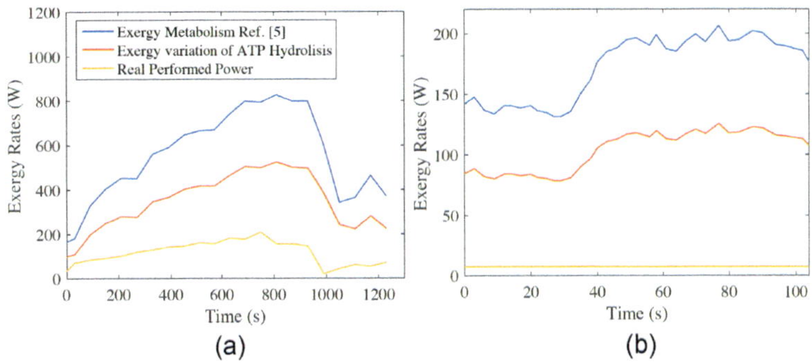

Fig. 4.5 Exergy terms indicating the metabolic efficiency, from the transformations of $\dot{B}_M$ into ATP ($\dot{W}_{MAX}$); and the conversion of ATP into $\dot{W}_{real}$ for a bicycle test. *Source* [3]

This indicates that glucose becomes the primary oxidized nutrient in the muscle cells for power generation.

Figure 4.7a displays a similar trend observed in the cycling test conducted on one male subject. The only point of inflection occurs when the impedance of the bicycle is suddenly reduced due to the cyclist's exhaustion, indicating the start of the resting period. The respiratory coefficient became close to one during the test, indicating that glucose is the primary form of energy obtained. However, in Fig. 4.7b, this ratio becomes higher than one, showing a clear distinction between the two activities. The prime activity (aerobic exercise) relies on aerobic respiration, while the latter has moments when anaerobic respiration is necessary to supply the energy demand

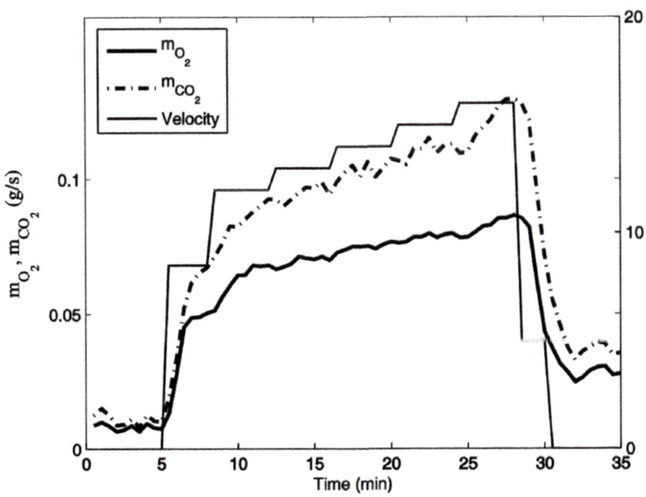

Fig. 4.6 Oxygen consumption rate and carbon dioxide production rate as a function of time for different exercise levels (velocity at left axis) for one subject. *Source* [4]

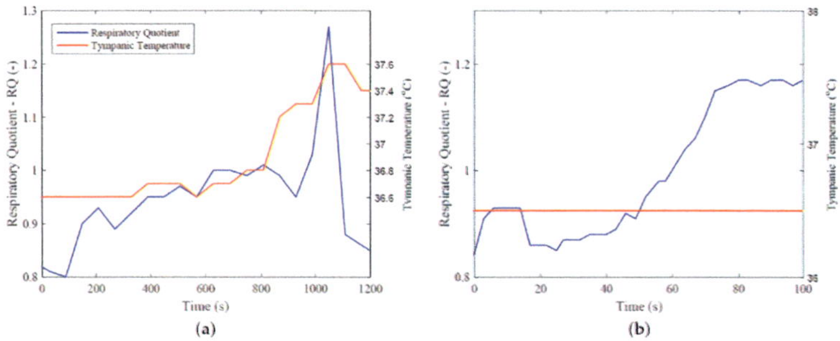

Fig. 4.7 **a** Respiratory quotient and tympanic temperature of a subject under a bicycle test, **b** respiratory quotient of a subject performing a continuous biceps series with 4 kg (the internal temperature was considered constant, equal to 36.5 °C). *Source* [5]

to produce work. The figure also includes the tympanic temperature measured over time, which will be further discussed in the following sections. For the biceps test, the tympanic temperature was considered constant, while in the bicycle test, the measurements were taken at every effort level.

With the data obtained from indirect calorimetry, including oxygen consumption ($\dot{m}_{O_2}$) and carbon dioxide production ($\dot{m}_{CO_2}$), it is possible to calculate the energy ($\dot{M}$) and exergy ($\dot{B}_M$) metabolism using Eqs. 3.27 and 3.28. In this process, it is important to differentiate between the heat transfer rate associated with the metabolic process ($\dot{Q}_M$) and the actual metabolism rate ($\dot{M}$), both measured in Watts. According to the first law of thermodynamics, for cellular metabolism, $\dot{Q}_M = \dot{M} - \dot{W}$. From the exergy analysis perspective, $\dot{B}_{dest} = \dot{B}_M - \dot{Q}_M \left(1 - T_0/T_{body}\right) - \dot{W}$. Hence, it is an error when the metabolism is treated as heat transfer (Joules) or heat transfer rate (Watts), something to bear in mind for the thermal comfort analysis [16]. From the perspective of exergy analysis, it is evident that these are different qualities of energy when transferred to their ability to perform work (Joules) or power (Watts). Figure 4.8 illustrates this concept, clearly showing that $\dot{Q}_M$ is in exergy basis ($\dot{B}_{\dot{Q}_M}$) lower than the performed power demonstrated in Fig. 4.4, for the same subject. Moreover, as expected for 'fuels' with simple chemical formulas, $\dot{B}_M \approx \dot{M}$.

The same reasoning can be applied to a subject in a bicycle test and during a weight lifting exercise. For simplicity, only one subject was analyzed for each test, allowing for a conceptual application of the equations. Therefore, oxygen and carbon dioxide consumption can be obtained from indirect calorimetry, as shown in Fig. 4.9a, b for each test. It is interesting to highlight that the carbon dioxide production follows a similar trend as the oxygen consumption for aerobic physical exercise (bicycle), as shown in Fig. 4.9a. However, for weight lifting, there is a point of tendency switching, where the mass production of CO_2 becomes even higher than O_2 consumption, indicating the moment that the activity becomes anaerobic, with fast burning of glucose, with no oxygen available for the body's total energy expenditure.

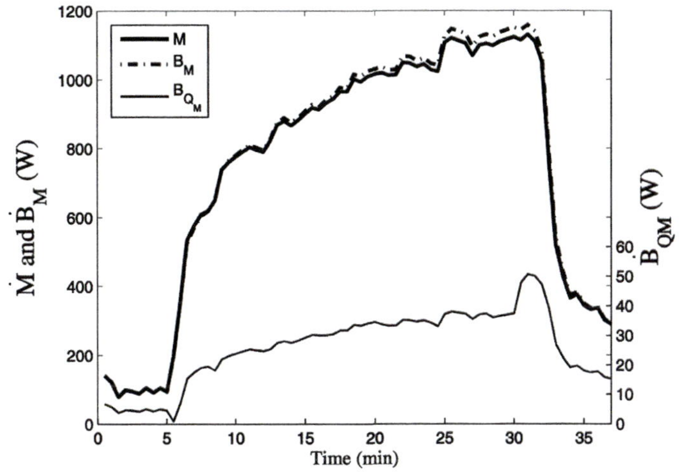

Fig. 4.8 Result of $\dot{M}$, $\dot{B}_M$ and $\dot{B}_{\dot{Q}_M}$ as a function of time during the treadmill test. The first two properties are indicated in the left axis, the last on in the right side axis. *Source* [4]

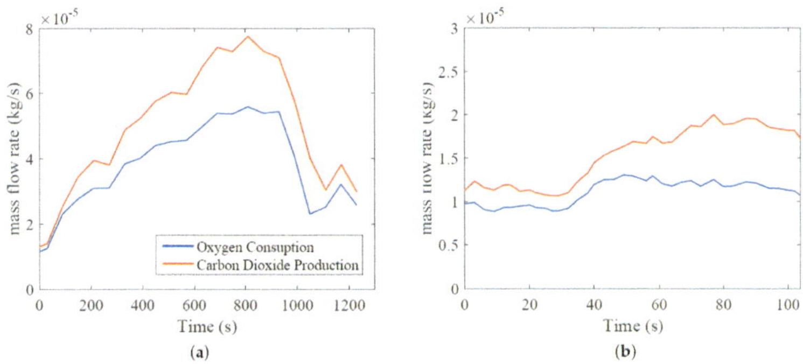

Fig. 4.9 Physiological data collected during tests: oxygen consumption and carbon dioxide production of a subject under: **a** bicycle test; **b** biceps series. *Source* [5]

From Eqs. 3.22 and 3.23 it is possible to calculate the consumption of glucose (representing carbohydrates) $\dot{m}_{carb}$ and lipids $\dot{m}_{lip}$, indicated in Fig. 4.10a, b. The consumption of lipids remains relatively constant throughout the test, while the oxidation of glucose increases significantly. This finding justifies the respiratory quotient being close to one, as the stoichiometry of the reaction is one-to-one for carbohydrates.

Based on the results of the indirect calorimetry, the metabolic energy and exergy for both tests were evaluated and compared with the values provided by the equipment. This is shown in Fig. 4.11a for the bicycle test and (b) for the weight lifting test. The comparison indicates that there is no significant difference in the values of $\dot{M}$ and $\dot{B}_M$. Moreover, it is evident that more intensive exercise leads to higher metabolism values, as seen when comparing the two types of exercise. The observed

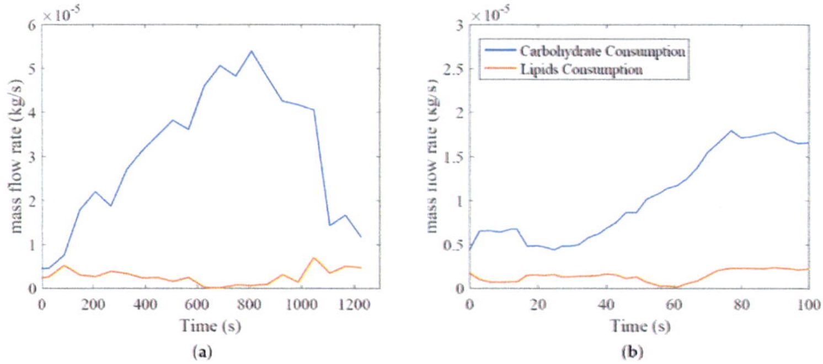

Fig. 4.10 Carbohydrate and lipid consumption during: **a** bicycle test; and **b** biceps curl. *Source* [5]

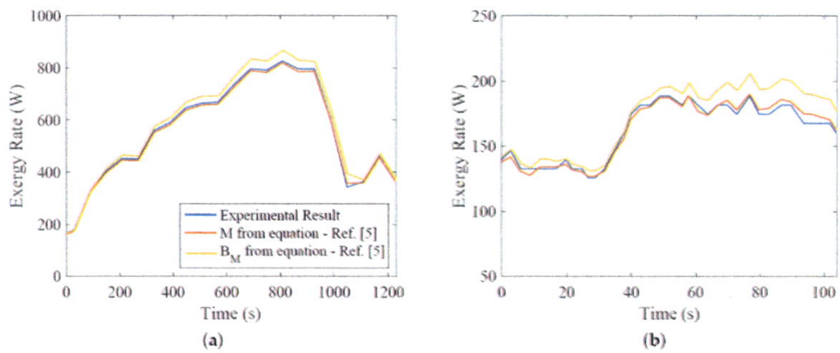

Fig. 4.11 Metabolism obtained from the equipment, calculated energy and exergy metabolism **a** Bicycle test; **b** weight lifting. *Source* [5]

difference in metabolism during the test was one order of magnitude, highlighting the exercises using larger musculoskeletal groups (legs and thigh muscles), whereas biceps tests use smaller muscles of the arm.

Using **test 2** as a model for comparing different participants, a proper statistical analysis was performed to calculate the average and standard deviation. The results are presented in Fig. 4.12. The subjects, including both males and females, performed the two activities on different days, which is a distinguishing feature of these results in relation to the literature. The same person performed two activities.

The metabolism, as measured by oxygen consumption and carbon dioxide production, is depicted in Fig. 4.12a, b. The results show a respiratory quotient (RQ) ratio close to one, indicating that the nutrient that is most used in these activities, treadmill and bicycle tests, is carbohydrates, represented by the smaller molecule glucose (used directly in the cells). Additionally, Fig. 4.12c, d display the metabolic rate and the esophageal temperature using Eq. 4.5 proposed by [17]. However, no similar equation has been found to provide insights into the internal body temperature during an exercise test. Moreover, there is a difficulty in measuring internal temperature during the exercise tests, although some digestible temperature pills are

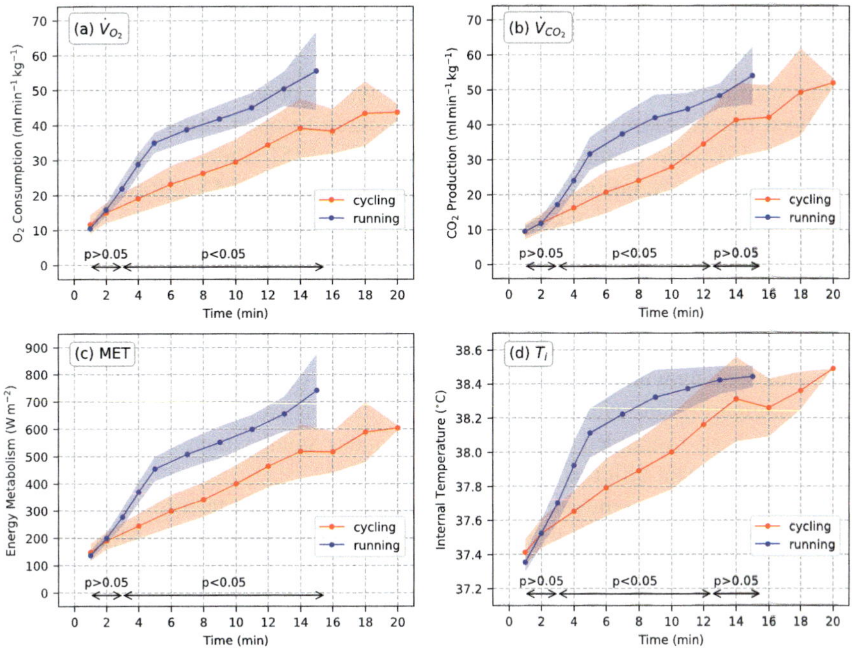

Fig. 4.12 Metabolism, oxygen consumption, carbon dioxide production and internal temperature of the body over time. *Source* [6]

already cited in academic studies such as the Bongers et al. [18] and may bring some new information on this research field.

$$T_{es} = 37.03 + 1.46 \left(\frac{\dot{v}_{O_2}}{\dot{v}_{O_{2,max}}} \right) \tag{4.5}$$

In Eq. 4.5, T_{es} represents the (esophagus) temperature, $\dot{v}_{O_2}$ represents the oxygen consumption at each moment t, and $\dot{v}_{O_{2,max}}$ represents the maximum oxygen consumption.

The increase in metabolism was higher during the running exercise, leading to a higher esophageal temperature. This indicates that the individual's body needed to release more energy into the environment. The intensification of energy transfer can be observed through increased intensity of the sweat evaporation and convection mechanisms (body parts movements) higher than running tests. These results will be further analyzed in the next sections.

4.2.3 Skin and Internal Temperature of the Body

In this section, we will focus on **test 2**, as the results obtained are more recent and had been treated more comprehensively compared to other studies conducted by the group [6], such as the data obtained in [1] and [5] where the skin measures of temperature were taken using an infrared thermometer (MX2, Raytek) at three locations: center of the back, posterior right arm, and posterior right leg. These measurements were taken during the last 2 min of each test speed. The results provide insights into the skin temperature profiles and behavior over time for cycling and running tests. For illustration purposes, Figs. 4.13 and 4.14 present the average skin temperature and their trends. The colored bands indicate the range of values between individuals, along with the standard deviation. The p-values resulting from unpaired t-tests comparing the cycling and running groups, as well as between males and females, are also included in the graphs. P-values smaller than 0.05 suggest that there are significant differences between the groups.

It is noteworthy that the same group of individuals running in ergometric (indoor) tests achieved lower surface temperatures compared to cycling tests. Although running tests are more energy-intensive for the body, the average skin temperature is lower than that of cycling tests (Fig. 4.13). One of the critical reasons is the need for increased heat dissipation to the environment due to higher oxygen consumption (metabolism), and the other reason is associated with greater movement of various body parts.

Figure 4.14 shows the skin temperature profiles for males and females during the test, where it is possible to infer that the temperature behavior is similar for both genders. These findings suggest that there are no significant differences observed in the temperatures of all limbs (p between 0.098 and 0.998 for all instants, $p > 0.05$); even though the data on the characteristics of the subjects (Table 4.1) indicate that there may be differences between genders in lean mass ($p = 0.034$), height ($p = 0.045$), and $\dot{v}_{O_2,max}$ in cycling ($p = 0.036$). Due to the similarity in temperatures among all individuals, the data from both cycling and running activities were analyzed as a combined group to compare the temperature differences between the two activities. The mean skin temperature data at each moment and the combined sample of all individuals indicate the hypothesis of normal distribution (Shapiro-Wilk test, between $0.066 \leq p \leq 0.85$ for all instants, $p > 0.05$) and homogeneity between the groups of each activity (Levene test, $0.054 \leq p \leq 0.41$ for all instants, $p > 0.05$).

With similar data from Fig. 4.12d, it was possible to obtain Fig. 4.15, where the subjects are only in cycling exercise test; therefore, with a known power performed over time. Figure 4.15 indicates the body's temperature as a function of time measured using the tympanic thermometer and applying Eq. 4.5, obtaining the internal temperature (esophagus) is possible. Moreover, by using the performed work, the internal temperature can be calculated based on the first law of thermodynamics ($dU_{\Delta T}/dt = m_{body}c_{body}dT_{body}/dt$), where the results show the same slope for the tympanic and esophagus temperatures. The difference in readings may be attributed to measurement inaccuracies of the tympanic, or such temperature being, in fact,

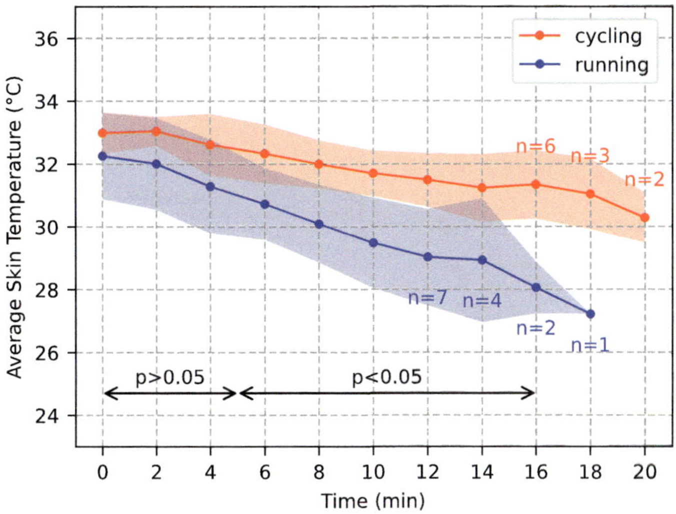

Fig. 4.13 Average skin temperature. The dots indicate the average values of the athletes at each speed, and the colored bands indicate the standard deviation between the subjects. The number of subjects that reached the last points is indicated in the graphs. The other points are composed of the whole sample, except for the following instants for males, due to data loss: running from 0 to 4 min ($n = 3$); cycling at 0 min ($n = 3$) and 2 min ($n = 2$). *Source* [6]

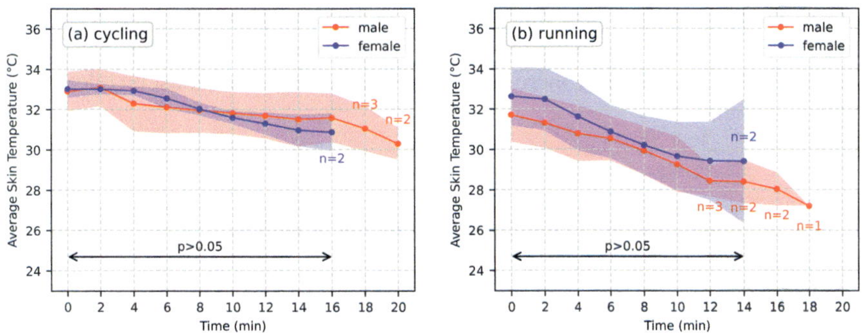

Fig. 4.14 Average skin temperature for males and females. The dots indicate the average values of the athletes at each speed, and the colored bands indicate the standard deviation between the subjects. *Source* [6]

lower than the esophagus one, or even potential errors associated with the equation. Additionally, it is noteworthy that when the maximum available work is used as an input in the first law of thermodynamics, it can lead to a decrease in temperature over time in certain cases; this observation raises the hypothesis that, in the absence of irreversibilities in the conversion of ATP's chemical power into work, the body's efficiency could be so high that it may result in a reduction in the core temperature.

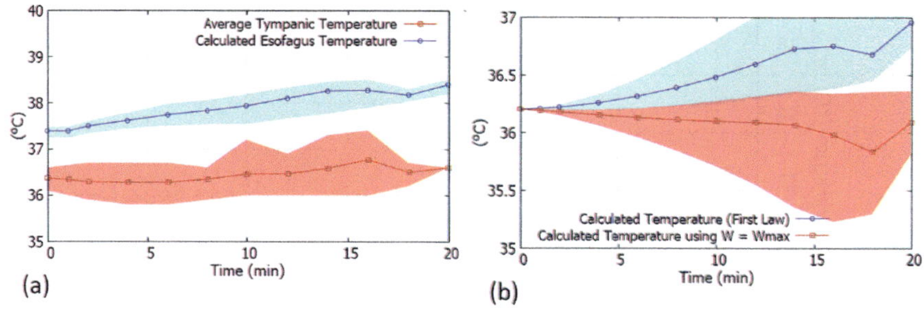

Fig. 4.15 Comparison of measured tympanic temperature, esophagus calculated temperature according to [17] **a**; and, application of the first law of thermodynamics with real performed power and application of the first law of thermodynamics with exergy variation of ATP hydrolysis, therefore no irreversibilities in conversion of ATP ($\dot{W}_{MAX}$) to Power ($\dot{W}$) **b**. Obs: The lines do not represent the pattern of the temperature behavior during the transition of the measured points. *Source* [3]

From the results of this section, it is possible to infer two trends. The skin temperature decreases over the exercise test, and the internal temperature increases. This result may be seen as contradictory at first sight, whereas a study of energy transfer of the body to the environment may give some clues. For instance, sweating is a mechanism to increase the enthalpy transfer rate of the skin to the environment, and it has the consequence of further reducing the skin temperature and raising the convection and radiation heat transfer rate. On the other hand, the body's internal temperature increases due to the increase of the metabolism, and all that energy can be seen as an internal energy increase of the body over time that must be released to the environment.

4.2.4 Energy and Exergy Transfer to the Environment

In Fig. 4.16, patients are composed of the least (subject 1) to the best trained (subject 11) runner. The separation was divided into three groups, based on the lactate threshold (LT) achievement speed. The first group (low LT) consists of four subjects with an LT speed lower than 3.1 m/s. The second group (medium LT) comprises three athletes with 3.1 *leq* LT speed *leq* 3.5 m/s. The final cluster (high LT) consists of four subjects with an LT speed higher than 3.5 m/s. During low-intensity, steady-state exercise, blood lactate concentrations do not increase beyond the normal biological variation observed at rest. As exercise intensity increases, blood lactate levels exceed the usual concentration. At a particular phase of the activity, blood lactate concentrations increase exponentially. Literature refers to [19] this inflection point as the lactate threshold (LT). In addition, we can use the lactate threshold to develop training programs to improve the performance of long-distance runners and triathletes.

For the **test 1**, Fig. 4.16a shows the calculated energy transfer rate (per surface area) to the environment through the skin (by convection, radiation, and evaporation) and the respiratory enthalpy variation through ventilation. The energy transfer increases fast at the beginning of the exercise due to increased evaporation (skin wetness) and ventilation. The main reasons for the increase in the energy transfer by those mechanisms are the increase in limb movements (which increase the heat transfer coefficient), the wet surface, and the respiratory tract flow. The saturation of the skin by sweat reduces the increase of the total energy transfer with the environment, which is observed in the graph (modification in the slope of the curve). Figure 4.16b shows the total exergy transfer rate (by surface area) to the environment. These exergy transfer's main mechanisms are related to enthalpy transfer (Fig. 4.16c, d). For all subjects, the maximum exergy transferred by convection is 2.1 W/m^2 and the maximum exergy transferred by radiation is 1.8 W/m^2. The evaporation through the skin Fig. 4.16c. The exergy transfer reaches a maximum value after the skin gets saturated by sweat. Figure 4.16d indicates the exergy variation by the pulmonary ventilation, which is proportional to metabolism and increases linearly as a function of speed. For basal conditions, the contributions of heat transfer rate and enthalpy flow rates to the environment during exercise are an order of magnitude higher than their exergy contribution. This indicates a first difference in applying the exergy analysis: the quality of the energy conversion process in the human body. Moreover, the person at basal conditions or rest does not perform any power and $\dot{M} = \dot{E}_{env}$, hence the energy released in the reactions of oxidation has the same magnitude as the energy transfer to the environment, but the nature is different. The exergy analysis only confirms this decrease in the quality of the energy by one order of magnitude.

Comparing Figs. 4.15 and 4.16, it is evident that there are challenges associated with evaluating the energy and exergy behavior of the human body during physical exercises. Proper calculation of mechanical work and power, accurate evaluation of the skin and internal temperature, are required. The application of the phenomenological model requires inputs such as convection heat transfer, surface wetness, and other factors that contribute to the complexity of the analyses. However, in the following Sect. 4.3, the exergy analysis is implemented in these tests, yielding promising results when compared to the indexes commonly used in the medical field.

4.3 Human Exergy Efficiency During Physical Exercises

In Fig. 4.17a, b, the destroyed exergy rate (per unit of mass) and the exergy efficiency are plotted as functions of the treadmill speed. It can be observed that the destroyed exergy rate increases with the increment of velocity, as oxidation reactions occur more rapidly, resulting in more irreversibilities. The exergy efficiency does not increase as sharply, which may be related to the growth of performed power at similar rates to the metabolic exergy. There is a slight increase in efficiency due to the increase in the exergy transfer to the environment. However, there is a saturation point where the skin becomes heavily covered with sweat, resulting in the destruction of exergy

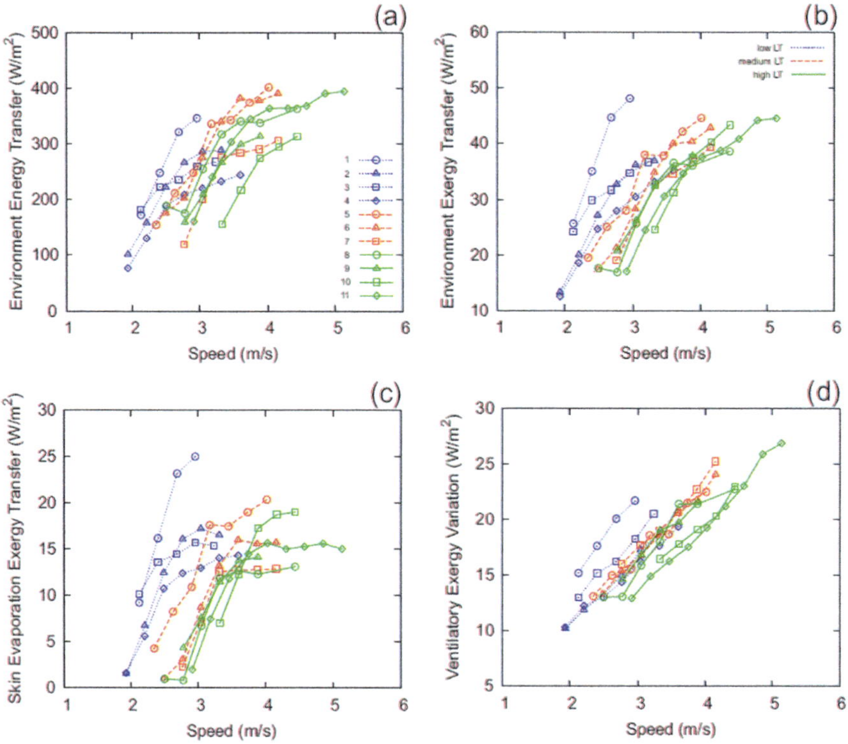

Fig. 4.16 **a** Total energy transfer rate to the environment through the skin ($\dot{E}_{env}/A_{skin}$), **b** total exergy transfer to the environment ($\dot{B}_{env}/A_{skin}$), **c** evaporative exergy transfer to the environment ($\dot{B}_e/A_{skin}$), **d** ventilatory exergy variation (($\dot{B}_{exp} - \dot{B}_{insp})/A_{skin}$. Mady et al. [1]

in the environment. Treadmill speed, where the values ranged between 30 and 44%. An unexpected result is that the subjects with low LT have higher exergy efficiency for the same speed, and their maximum occurs for lower speed compared to others. This last result can be explained by the exergy transfer to the environment, primarily through evaporation). The exercise for this group began at a lower speed; their skin gets saturated with sweat earlier.

Figure 4.18 provides a summary of the previous figures repeating Fig. 3.5. It showcases the efficiency of energy conversion, the exergy transfer associated with heat, and the irreversibilities in each metabolic pathway (representing the degradation of nutrients) to perform work in a given activity. The significance of the results in Fig. 4.18a, b is that, instead of evaluating different performed powers, the main objective was to choose the variables to measure and integrate them over time.

Column (1) indicates the maximum work the body could extract from nutrients referred to as metabolic exergy (in blue). The integration of this physical quantity over the test duration resulted in the value obtained in Fig. 4.18a being 730 kJ, and for Fig. 4.18b being 18 kJ. Column (2) shows that the production of ATP involves a complex chain of reactions with multiple irreversibilities occurring at each step of combining (ADP + P_i). There is an intrinsic destroyed exergy in the process, represented in red ($\dot{B}_{d,ATP}$), and an exergy loss associated with heat, represented in green ($\dot{B}_{Q,ATP}$). In the last step, column (3), the energy conversion process from ATP (hydrolysis of this substance) to performed power is shown in blue, with its exergy lost as heat ($\dot{B}_{Q,r}$) and destroyed ($\dot{B}_{d,r}$). Only $\dot{B}_{available}$ represents the exergy used to perform the physical activity as work ($\dot{W}$), and it is higher for the bicycle test than for weight lifting.

Figure 4.19 shows the exergy efficiency as a function of maximum oxygen consumption per muscle mass. This physical property ranged from 23 to 28% efficiency, which is the first time the group obtained this physical property based on real performed work and not solely on a different set of equations. The reason for this figure is the argument that for subjects running on a treadmill, different equations to obtain the performed power in the literature lead to values ranging from 343 to 1650 W, for speeds between 3.6 and 3.9 m/s, as discussed earlier [7–9, 12]. In Fig. 4.19, it was possible to establish this efficiency since the performed power is a known variable.

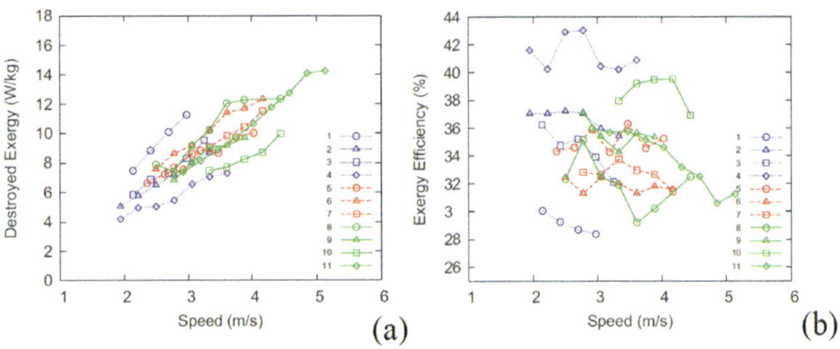

Fig. 4.17 **a** Destroyed exergy rate (W/kg) and **b** Exergy efficiency during the incremental test evaluation. *Source* [1]

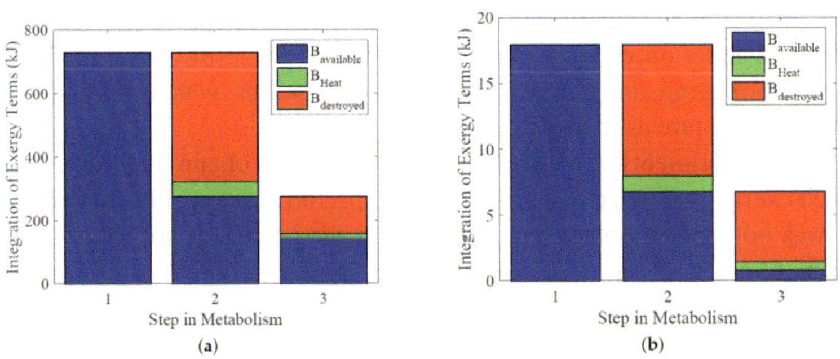

Fig. 4.18 Exergy conversion process in cellular metabolism. From nutrient oxidation, ATP formation and the use of ATP. Based on the figures proposed in: **a** Bicycle test; **b** weight lifting. *Source* [5]

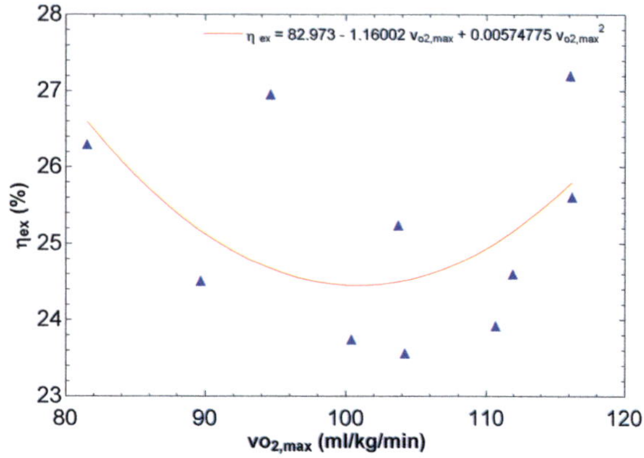

Fig. 4.19 Exergy efficiency as a function of maximum oxygen consumption per muscle mass ($R^2 = 30\%$). *Source* [3]

Figure 4.19 highlights the importance of properly evaluating the phenomenological model of the human body before applying the second law of thermodynamics. There are several limitations to overcome in the phenomenology of the model; for instance, refining the heat transfer coefficient by convection and improving the evaluation of skin wetness. As shown in Fig. 4.17, the heat and associated exergy with this heat influence the body's exergy efficiency, as it is the term responsible for maintaining the body at a certain temperature. The wetness factor is particularly important for maintaining homeothermy. Regarding the exercise, the group's next steps should include a thorough analysis of the biomechanics of the body. However, it is worth noting that most equations used in the study did not violate the second law of thermodynamics. Despite the limitations, the group was able to assess various variables and determine that the bicycle test is the most suitable for evaluating the second law of the human body and assessing the quality of energy conversion in human body metabolism.

For a person in resting conditions, several authors of literature obtained that living beings tend to a minimum entropy production rate. Balmer [20] and Aoki [21] obtained these results from experimental analyses, the former for fishes and the latter for humans. The results of Mady et al. [22] show that the destroyed exergy decreases as a function of aging, and the exergy efficiency decreases, which was at that time a distinguishing feature as indicated by Fig. 4.20.

From the experiments of the **test 1**, it was possible to obtain that runners had the destroyed exergy rate and metabolism rate and performed power integrated over the test. It was possible to obtain Fig. 4.21, where the exergy destroyed decreased as a function of aging. What is interesting is that the exergy efficiency increases with aging. Several hypotheses may be raised as an adaptation of the body to that kind of activity, but no conclusive explanation was found.

Some simulations were conducted in **tests 1 and 2** to evaluate different ratios of W/W_{MAX}, since the performed power is an unknown variable for runners.The purpose is to find some correlation between W/W_{MAX} in relation to B_{dest}/A_d.

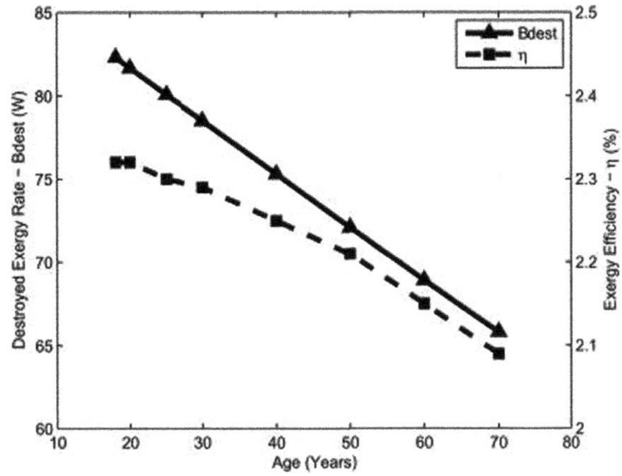

Fig. 4.20 **a** Ratio of the destroyed exergy and metabolism (in exergy basis) and **b** exergy efficiency as a function of the subject age. *Source* [22]

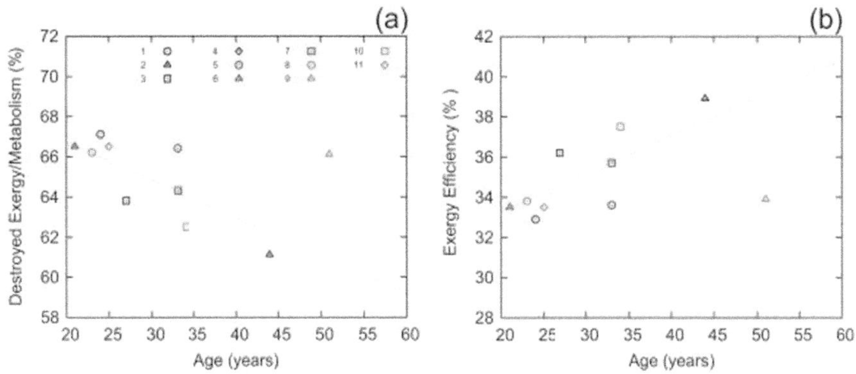

Fig. 4.21 Ratio of performed work with maximum available work as a function of: **a** destroyed exergy; **b** exergy transfer to environment associated with metabolism. *Source* [22]

To obtain Fig. 4.22, the exergy terms were integrated during the activity and divided by the body surface area and test duration (average normalized values [J/(m^{2}s)]), with an aim to compare all runners on the same basis. These results suggest that the global exergy efficiency is around 10 to 50% with ratios of W/W_{MAX} of 6 to 30%. It is interesting to highlight that as in [1] for the same destroyed exergy, better trained subjects (or subjects with higher maximal oxygen consumption) performed a higher power, Fig. 4.22b or used the exergy content in ATP more efficiently Fig. 4.22a. The only tendency that was different was to runner 2; in the future, more female subjects must be chosen for the test in order to obtain better conclusions regarding this difference between male and female from the Second Law optics.

Finally, when the integrated values of destroyed exergy and exergy transfer associated with metabolism [1], are transformed into an average rate over time, the filled

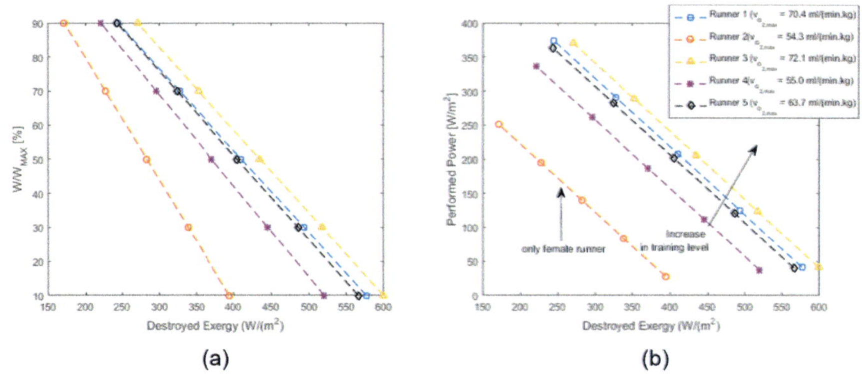

Fig. 4.22 **a** Ratio of the destroyed exergy and metabolism (in exergy basis) and **b** exergy efficiency as a function of the subject age. *Source* [23]

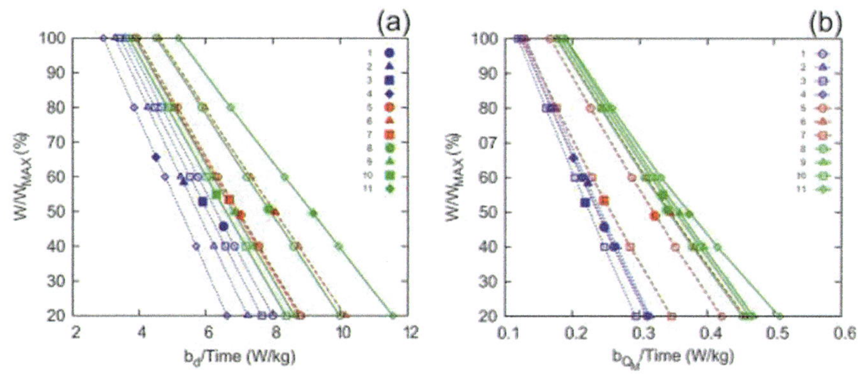

Fig. 4.23 Destroyed exergy for different values of performed power and different ratios of $\dot{W}$ to $\dot{W}_{Max}$. *Source* [1]

dots in Fig. 4.23 are obtained. The lines with unfilled dots correspond to the exergy analysis applied to each subject for different conditions of W/W_{MAX} (relation of work actually performed to the maximum available work), but for the same collected data. The purpose of the analysis was to correlate different ratios of W/W_{MAX} with two exergy terms: destroyed exergy per subject mass (b_d) and exergy transfer per subject mass (b_{Q_M}) associated with metabolic heat transfer. Moreover, whereas all previous results did not indicate any clear trend as a function of the functional capacity, Fig. 4.23 indicates a clear discrimination. For the same destroyed exergy the better trained subjects have higher ratios of W/W_{MAX}. Figure 4.23b leads to similar conclusions, therefore, for the same exergy transfer bQM, the better trained subjects can perform more work. Another curious discussion is that for the same ratio W/WMAx the better trained subjects destroy more exergy per subject mass and dissipate more exergy as heat per subject mass.

4.4 Concluding Remarks

In this chapter, a state-of-the-art analysis of the energy and exergy behavior of the human body during physical exercise was conducted, which allowed for the assessment of efficiencies for cyclists at approximately 30%. Nevertheless, the model reveals several limitations for future analyses, such as determining the actual power output during running and weightlifting. How does the sweat change along the test? Does the body get saturated (wetness $\omega + d\omega \approx 1$)? Nevertheless, some results show the potential of exergy analysis as a tool for proposing indicators to support research in the area with quality indicators.

The body temperature tends to increase as a function of treadmill speed while the skin temperature remains almost constant. The first is more dependent on physiological variables, whereas the second is more dependent on environmental variables.

He destroyed exergy rate increases as a function of speed, as previously obtained in the literature, but for entropy production. The ratio of the destroyed exergy to metabolic exergy remains almost constant as a function of speed, but it tends to decrease as a function of the subject's age, as occurs for basal conditions (Prigogine principle). The exergy efficiency tends to increase with age, which differs from the basal conditions' results.

The metabolic path, from nutrients' consumption (obtained by indirect calorimetry) to performed power, was first analyzed from the exergy analysis point of view. The exergy efficiency achieved values around 40% if the exergy input considered was the ATP and values around 30% for bicycle tests, whereas the efficiency was no larger than 10% for the weight lifting.

References

1. Mady CEK, Albuquerque C, Fernandes TL, Hernandez AJ, Saldiva PHN, Yanagihara JI, Oliveira-Junior S (2013) Exergy performance of human body under physical activities. Energy 62:370–378
2. Mountcastle VB (1974) Medical physiology. CV Mosby
3. Mady CEK, Igarashi TL, Albuquerque C, Santos-Silva PR, Fernandes TL, Hernandez AJ (2019) Exergy efficiency on incremental stationary bicycle test: A new indicator of exercise performance? J Brazilian Soc Mech Sci Eng 41:1–11
4. Mady CEK, Oliveira-Junior S (2013) Human body exergy metabolism. Int J Thermodyn 16(2):73–80
5. Spanghero GM, Albuquerque C, Lazzaretti FT, Hernandez AJ, Mady CEK (2018) Exergy analysis of the musculoskeletal system efficiency during aerobic and anaerobic activities. Entropy 20(2):119
6. Igarashi TL, Fernandes TL, Hernandez AJ, Mady CEK, Albuquerque C (2022) Behavior of skin temperature during incremental cycling and running indoor exercises. Heliyon 8:10
7. Borg G (1973) Psychophysical scaling with applications in physical work and the perception of exertion. Med Sci Sports Exerc 5(2):90–93
8. Kaneko M (1990) Mechanics and energetics in running with special reference to efficiency. J Biomech 23:57–63

9. Ito A, Komi PV, Sjödin B, Bosco C, Karlsson J (1983) Mechanical efficiency of positive work in running at different speeds. Med Sci Sports Exer 15(4):299–308
10. Williams KR, Cavanagh PR (1983) A model for the calculation of mechanical power during distance running. J Biomech 16(2):115–128
11. Bijker KE, Groot G, Hollander AP (2001) Delta efficiencies of running and cycling. Med Sci Sports Exer 33(9):1546–1551
12. Cavagna GA, Kaneko M (1977) Mechanical work and efficiency in level walking and running. J Physiol 268(2):467–481
13. Powers SK, Howley ET (2000) Fisiologia do exercício: teoria e aplicação ao condicionamento e ao desempenho. Manole
14. Lems S (2009) Thermodynamic explorations into sustainable energy conversion. Learning from living systems. PhD thesis, Technische Universiteit Delft
15. Cavagna GA (1975) Force platforms as ergometers. J Appl Physiol 39(1):174–179
16. Ribeiro TJS, Mady CEK (2022) Comparison among exergy analysis methods applied to a human body thermal model. Energy 239:122446
17. Nadel ER, Bullard RW, Stolwijk JA (1971) Importance of skin temperature in the regulation of sweating. J Appl Physiol 31(1):80–87
18. Bongers CC, Hopman MT, Eijsvogels TM (2015) Using an ingestible telemetric temperature pill to assess gastrointestinal temperature during exercise. J Vis Exp: JoVE 104:53258
19. Fernandes TL, Hernandez AJ, Albuquerque C, Mady CEK (2018) Aplicação clínica da termografia para equilíbrio energético em atletas-estado da arte e novos paradigmas. Revista Brasileira de Medicina do Esporte 24:483–485
20. Balmer RT (1982) Entropy and aging in biological systems. Chem Eng Commun 17:171–181
21. Aoki I (1994) Entropy production in human life span: A thermodynamical measure for aging. Age 17(1):29–31
22. Mady CEK, Ferreira MS, Yanagihara JI, Saldiva PHN, Oliveira-Junior S (2012) Modeling the exergy behavior of human body. Energy 45(1):546–553
23. Igarashi TL, Spanghero GM, Fernandes TL, Hernandez AJ, Mady CEK, Albuquerque C (2018) Exergy analysis of the body efficiency during aerobic activities. In: Proceedings of ecos 2018–the 31st international conference on efficiency, cost, optimization, simulation and environmental impact of energy systems june 17–22, Guimarães, Portugal

Chapter 5
Exergy Analysis and Thermal Comfort Conditions

5.1 Human Thermal Model–Phenomenological Behavior of the Body

A comprehensive model of the human body's thermal system, initially proposed by [1] was utilized and subsequently modified (simplified) for the purpose of applying exergy analysis to different anatomies [2–4]. Initially, a simplification is proposed to address cases where there is a significant and challenging increase in body mass, such as slim and fat individuals. Implementing such changes in a complex model can be difficult. These modifications led to the works of [2] and [3]. Another application focused on the utilization of the second law of thermodynamics methods applied to the human thermal model proposed by [1]. This research was conducted by [6], who discussed and compared various methods of applying exergy analysis using the [1] model as a basis. The result shows that the method proposed by [5] is more sensible to several relative humidities and leads to values that are closer to metabolism.

Figure 5.1 depicts the human thermal model used by [2–4]. The [1] model employs the following parameters for the human body: height 1.76 m, mass 67 kg, body area 1.8 m^2, and volume 0.00627 m^3. To represent the human body in the model, it utilizes 15 cylinders of elliptical cross–sections. Additionally, each body segment in the model is composed of a combination of the following tissues: skin, fat, muscle, bone, brain, viscera, lung, and heart. The energy balance is solved individually for each tissue, which is assigned its respective metabolism rate, thermophysical properties, and blood perfusion characteristics. The heat exchange between the blood and the tissue is taken into account in the energy balance equation for each element. The vessels are categorized into two types: large and small. The small vessels were treated as part of a continuum, following the model proposed by [7]; later, the model was adjusted and validated by [8]. However, the blood temperature within each tissue is influenced by the location within the model. The treatment of large vessels follows the approach proposed by [9]. In the model, each tissue element is equipped with

C. E. Keutenedjian Mady, *Exergy Analysis*, Green Energy and Technology,
https://doi.org/10.1007/978-3-032-01646-1_5

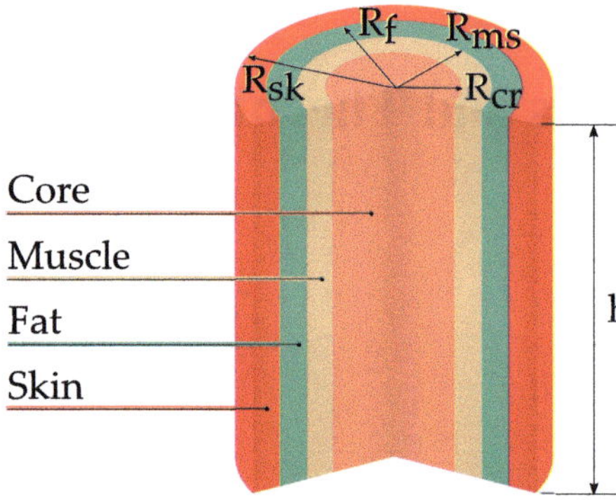

Fig. 5.1 Human thermal model used in this chapter to assess thermal comfort conditions, where it is indicated each boundary condition. The only part not presented is the central reservoir of blood. *Source* [2–4]

two blood reservoirs: one for arterial blood and the other for venous blood. These reservoirs facilitate heat exchange between each other, as well as with the adjacent tissues and elements. Notably, the trunk region comprises only one blood reservoir instead of two.

The initial geometry of the body was calculated using the data for the standard man and was based on [1]. Werner and Buse [10] present some physical and thermal properties of many organs and tissues' volumes, where [2] obtained Table 5.1.

Since the human thermal model body is a cylinder, it is impossible to keep all the parameters of the standard man. Tests made with the model showed that the temperature profile is closer to reality if the height of the standard man is maintained instead of its surface area. Therefore, if the cylinder maintains the original human area, the results are not viable heights, or if the original height is maintained, the surface area suffers some distortion. This approach could lead to values of heat transfer that are not similar to reality. At this point, if we work with absolute values in Watts or Joules, the heat transfer to the environment is the same for a person with those characteristics; nevertheless, normalizing using the area of the model may lead to mistakes; the correct path is to normalize using DuBois [12] area (W/m^2) the results are close to the expected in experiments. Moreover, the two-node model proposed by Gagge [11] is a reference even in thermal comfort standards.

Table 5.1 Thermophysical properties of the layers. Based on literature [2]

Layers	ρ (kg/m^3)	k (W/(m.K))	c (J/(kgK))	M''' (W/m^3)	$\dot{\omega}_{bl_t}$ (m$_s^3$/(m$_t^3$s))
Skin	1085	0.43	3680	368.1	361.7
Fat	920	0.21	2300	368.4	76.7
Muscle	1085	0.51	3800	684.2	542
Core	1119	0.5038	2679	2653	4155

Table 5.2 Dimensions and volumes of each layer for the original anthropometric data, obtained by [2]

Layer	Mass (kg)	Volume (cm^3)	External radius (cm)
Skin	3.3246	3287	10.54
Fat	8.508	10153	10.26
Muscle	27.36	27688	9.326
Core	20.78	20426	6.078
Blood	5.825	5500	–

Most of the human body blood is distributed along tissues and organs in the data provided by Werner and Buse [10]. Considering that the volume of this element is the same as provided by the authors, their mass must be corrected. The correction is made by estimating the total blood volume using the equation proposed by Nadler [13], and considering that this blood is equally distributed along all organs and tissues. The volume of the layers is calculated by adding the volume of each element that composes it. Note that, after the correction, the blood volume is disregarded for the body volume calculation, but its mass should be taken into account. Table 5.2 demonstrates these weightings and the results of each tissue thermophysical property.

The volume and height of the anatomical model were maintained. In this way, obtaining the external radius values was possible by successively subtracting the volumes of each layer. To change the anatomical conditions such as metabolism, mass, and height, it was necessary to calculate the surface area to obtain the body volume from the external radius. When the normalization is conducted by surface area, the same technique used in [20] was employed, the division by the [12]. For each anatomical configuration, the basal metabolism and the blood volume can be calculated using Eqs. 5.1 and 5.2. This last equation was proposed by [13].

$$\dot{M}_{basal,neutrality} = \dot{M}_s''' V_s + \dot{M}_f''' V_f + \dot{M}_m''' V_m + \dot{M}_c''' V_c \tag{5.1}$$

$$V_b[m^3] = (0.3669h^3 + 0.03219m_{boby} + 0.6041)/1000 \tag{5.2}$$

This procedure was conducted for each anatomy, where, depending on the increase or decrease of lean or fat mass, it was necessary to modify the configuration of the model. Moreover, the volume of blood was adjusted in accordance with specific anatomy. Table 5.2 provides the reference anatomy that was considered.

5.1.1 Energy Balance of the Human Thermal System

To obtain the temperature distribution within the model, the first law of thermodynamics must be solved, as indicated in Eq. 5.3. For blood perfusion, the [7] was adopted. The subscript t refers to different tissues (skin, fat, muscle, and core), and *bl* represents the blood. The main assumptions of the model are as follows: (i) each layer has properties uniformly distributed throughout the volume; (ii) the temperature within each layer is considered constant, following the concentrated capacitance method; it is considered the axial direction of the cylinder, heat conduction; arterial blood has the same temperature (time-dependent); the blood leaving the tissues is in thermal equilibrium with them; in some simulations, the model is in the anatomical position (basal, sedentary, office conditions).

$$\rho_t c_t V_t \frac{\partial T_t}{\partial t} = \frac{\partial}{\partial r}\left(k_t A_t \frac{\partial T_t}{\partial r}\right) + (\rho_t V_t)(\rho_s c_s)\dot{\omega}(T_s - T_i) + M''' V_t \tag{5.3}$$

Note that boundary conditions at the skin involve energy exchanges with the environment, including convection, radiation, evaporation, and respiration. It is important to mention that the enthalpy change associated with respiration is considered only in the core. Under the assumption of concentrated capacitance, the term $\frac{\partial}{\partial r}\left(k_t A_t \frac{\partial T_t}{\partial r}\right)$ simplifies to a difference between the heat received by conduction from the previous tissue and the heat lost by conduction to the next tissue. However, it is important to highlight that the skin has a special case, as it directly exchanges energy with the environment. These relationships are depicted in Eqs. 5.5–5.8.

In these equations, the term, $\dot{Q}_{t\rightarrow t+1}$, represents the conduction heat transfer rate between each tissue (t) and the one in contact ($t+1$), e.g., from fat to skin. The energy transfer between blood and each tissue t is calculated using $\Delta\dot{H}_{bl\rightarrow t}$, similar to the central reservoir. The skin has three terms related to energy transfer rate: radiation, convection, and the diffusion of water and subsequent evaporation through the skin. Finally, the energy transfer associated with respiration, including the increase in air humidity and temperature, is concentrated in the core region. Equation 5.4 quantifies the absolute value of the heat transfer through conduction between a more central layer (t) and the next ($t+1$).

$$\dot{Q}_{t\rightarrow t+1} = \frac{2 \times \pi \times k_{eq}}{ln\left(\frac{r_{t+1}}{r_t}\right)} \times (T_t - T_{t+1}) \tag{5.4}$$

$$\rho_c \forall_c c_c \frac{dT_c}{dt} = -\dot{Q}_{c\rightarrow m} + \Delta\dot{H}_{bl\rightarrow c} + \dot{M}_c''' \forall_c - \Delta\dot{H}_{resp} \tag{5.5}$$

$$\rho_m \forall_m c_{p,m} \frac{dT_m}{dt} = \dot{Q}_{c\rightarrow m} - \dot{Q}_{m\rightarrow f} + \Delta\dot{H}_{bl\rightarrow m} + \dot{M}_m''' \forall_m \tag{5.6}$$

$$\rho_f \forall_f c_f \frac{dT_f}{dt} = \dot{Q}_{m\rightarrow f} - \dot{Q}_{f\rightarrow s} + \Delta\dot{H}_{bl\rightarrow f} + \dot{M}_f''' \forall_f \tag{5.7}$$

$$\rho_s \forall_s c_s \frac{dT_s}{dt} = \dot{Q}_{f\rightarrow s} + \Delta\dot{H}_{bl\rightarrow s} + \dot{M}_s''' \forall_s - \dot{Q}_{conv} - \dot{Q}_{rad} - \dot{H}_e \tag{5.8}$$

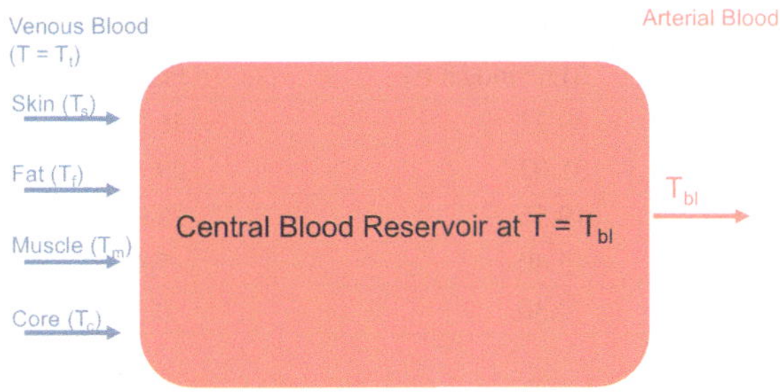

Fig. 5.2 Central reservoir of blood receiving venous blood (T_t) from several tissues (t) and dispatching arterial blood (T_{bl}) leaving to all tissues. *Source* [2–4]

Finally, the first law of thermodynamics is applied to the central blood reservoir (Fig. 5.2), accounting for the energy exchange between blood and tissue (t = skin, fat, muscle and core)–$\Delta\dot{H}_{bl\to t}$. This term represents the energy balance within the central reservoir, which receives blood from the tissues and delivers blood to the tissues at arterial blood temperature. The central reservoir serves as a representation of the large vessels and tissues, providing a simplified approach, since the heat exchange between both reservoirs is not taken into account. Equation 5.9 describes the energy balance within the blood reservoir.

$$\rho_{bl}\forall_{bl}c_{bl}\frac{dT_{bl}}{dt} = (\rho_{bl}c_{bl})\sum_{i=1}^{n=4}(\rho_{t,i}\forall_{t,i})\dot{\omega}_{b_{t,i}}(T_{t,i} - T_s). \tag{5.9}$$

5.1.1.1 Sex Differences in the Models

As Kyle et al. [14] indicated, males and females have different amounts of fat and muscle mass. The data presented by Werner and Buse [10] only refer to males; therefore, a literature review of body composition was conducted using previous studies by Kyle et al. [14], Janssen et al. [15], and Flegal et al. [16]. The composition displayed in Kyle et al. [14] (males and females) was selected to represent the skeletal and smooth muscles of the models. Moreover, data from this study were used for the model compositions of males and females, as shown in Table 5.3.

Another significant difference between males and females in office environments is the type of clothing and their clothing related requirements. Typically, men wear thicker clothing, such as suits. In this study, typical men's and women's office clothing is considered: a full suit for the male model (with a clothing insulation of $I_{ml} =$ 0.96 CLO) and a skirt and long shirt for the female model ($I_{fem} =$ 0.67 CLO) [17].

Table 5.3 Mass percentage of each layer in the male and female models (based on [14])

Layer	Male model (%)	Female model (%)
Core	39.73	36.58
Muscle	35.02	29.28
Fat	19.55	28.44
Skin	5.40	5.40
Blood	0.30	0.30

5.1.1.2 Menstrual Cycle

The female hormonal cycle lasts 28 days on average and can be divided into three main stages. The beginning of the cycle is considered to occur on the first day of menstruation. The maturation of the ovarian follicle commences on this day. The growth of follicles causes the discharge of the hormone estrogen, which peaks approximately in the middle of the period (Fig. 5.3), and the high concentration of estrogen incites ovulation. After ovulation, the luteal phase commences, which lasts approximately 14 days and is marked by an increase in the discharge of the hormones, estrogen and progesterone. By the end of the luteal phase, there is a decrease in the levels of these two hormones, resulting in menstruation and the beginning of a new cycle, as explained by Hema Latha [25].

There are two types of approaches that are used in this research area: (i) studies that indicate a correlation between metabolic variations and the different phases of the menstrual cycle; and (ii) studies that claim that there is no proven relationship.

Hema Latha [25] conducted measurements of $\dot{v}_{O_2}$ and the $\dot{M}_{basal}$ throughout different stages of the female hormonal cycle. The results indicated an important development in the $\dot{M}_{basal}$ during the luteal phase. However, the variation between the menstrual and follicular stages was not significant. Additionally, Frascarolo et al. [27] studied thermal regulation and thermoregulatory mechanisms and found no change in metabolism during menstrual cycle phases. Solomon et al. [28] also suggested an increase in metabolic rates during the luteal phase compared to the menstrual period. They noted low levels of BMR in the preovulatory period when

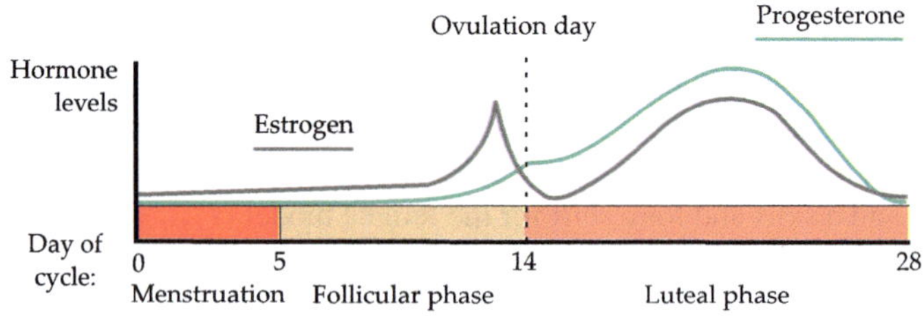

Fig. 5.3 Hormone concentrations during the menstrual cycle. Adapted from Slauterbeck et al. [26]

the estrogen concentration is high, which indicates that the BMR does not dominantly affect metabolic regulation. Other studies have indicated an increase in metabolic activity during the luteal phase. In this study, an increase of metabolism activity of 13.4% during the luteal phase was utilized, as obtained from the average values reported by Bisdee et al. [29] and Hema Latha [25].

5.1.1.3 Thermal Comfort Indicators

The thermal activity index (TAI), used as an indicator of energy variation in the body over time, measures the difference between the energy associated with metabolism for a class of activity and the energy exchanged by skin and respiration under neutral thermal conditions for the same activity and environment. Fanger [17] subjected the volunteers to different thermal environments and asked them to rank thermal sensations from very cold (–3) to very hot (+3), and the thermal neutrality, or thermal comfort in some conditions being 0. From a linear regression of these experimental data as a function of $\dot{M}$" and TAI, the equation for the predicted mean vote (PMV) was obtained. This indicator estimates people's thermal sensation of environments. The percentage of people dissatisfied (PPD) term indicates the thermal discomfort experienced in the thermal environment. Equations 5.10 and 5.11 are used to calculate the PMV and PPD, respectively.

$$PMV = \left[0.303 \exp\left(-0.036\,\dot{M}''\right) + 0.0028\right] TAI \tag{5.10}$$

$$PPD = 100 - 95 \exp\left(-0.003353\,PMV^4 + 0.2179\,PMV^2\right) \tag{5.11}$$

5.1.2 Thermal Load Evaluation

This section details the Cooling Load Temperature Difference (CLTD) method for calculating thermal loads in environments. The article by Spitler et al. [19] provides a review of this method, proposed by the American Society of Heating, Refrigerating and Air-Conditioning Engineers (ASHRAE). According to Spitler et al. [19], this method for calculating thermal loads in environments was first proposed in the Cooling and Heating Load Calculation Manual [20], suggesting a means to estimate heat transfers between the environment and the surroundings, as well as heat generation within the environment (due to occupants and equipment). Cooling loads are calculated for each hour of the day, then they are added together to obtain the daily load or they are evaluated at the peak value. The procedures outlined by the CLTD method for calculating thermal loads are detailed for outdoor and indoor walls, ceilings, occupants, equipment, and glazing (or fenestrations), as depicted in Fig. 5.4.

Past work has aimed to combine minimizing exergy analysis and the building envelope [21], and the main concepts are summarized in [22]. Nevertheless, when

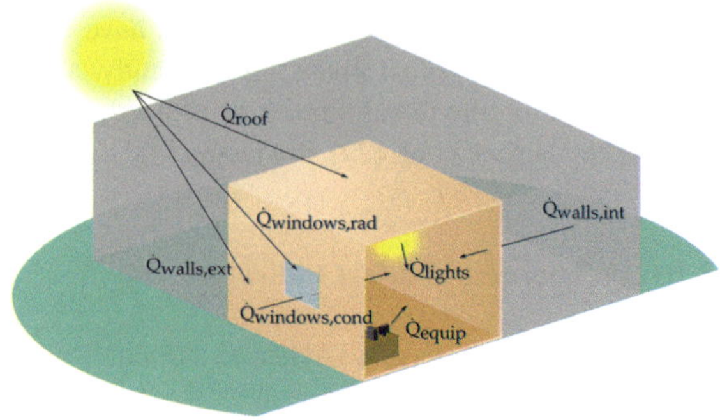

Fig. 5.4 Schematic sketch showing the thermal exchanges of one modeled environment (in beige) with its exterior and adjacent rooms (in gray)

there is a combination of males and females in the same thermal environment, the proposition of the best resistance (amount of clothes) to improve the thermal comfort index established by [6] has not yet been investigated in the literature. After the results of [3], some studies were found in the literature [23, 24]. However, there is still little emphasis in scientific production on the exclusive analysis of the female body with all the indexes proposed. Given that women represent more than 50% of the potential active workforce, the importance of promoting research focused on the female anatomy becomes evident, aiming for more significant equity in thermal sensation, bearing in mind that the following consequence is to reduce building energy consumption. The main objective is to show that the occupants of the thermal environment need to change their clothes, increasing thermal resistance due to a barrier to air temperature. Furthermore, there must also be further investigation in order to achieve a higher number of satisfied people with lower energy demands, as discussed by Molliet and Mady [3, 4].

Moreover, we used an actual thermal environment in São Paulo, Brazil, to simulate occupational behaviors and differences, as indicated in Fig. 5.5. The thermal environment modeled is based on the Philadelphia Building, a commercial building located in the district of Higienópolis, in the city and state of São Paulo. The building consists of 15 floors with commercial units, a bank branch on the ground floor, and a convention hall, whose division of main areas is shown in Table 5.4. The Philadelphia Building was chosen as the object of study because it was easy to obtain the parameters. The floor plans of the commercial units are standardized according to Fig. 5.6. The ceiling height of each floor is 2.6 m. To simplify the modeling, the faces of the building were assumed to be parallel to the cardinal directions (although this does not correspond to the reality of the building; such information would not significantly alter the result, but it would render the use of the tables in [20] almost impractical).

For the calculation of thermal loads associated with the walls and ceiling, the dimensions and parameters from Table 5.5 were used. The areas were obtained from the floor plan in Fig. 5.6 and the ceiling height. The adjustment factor LM is tabulated and adjusts the data from the CLTD tables (whose standard is the Northern Hemisphere) for Southern latitudes (given the building's location in the city of São

Fig. 5.5 Photo of the main facade of the Philadelphia Building. *Source* [3, 4]

Table 5.4 Division of the building envelope. *Source* [3, 4]

Environment	Private area [m^2]
Commercial unities	324
Bank branch	309
Convention room	122

Paulo). The month of December, typically a warm period in São Paulo, was chosen. The wall areas for each face have already excluded the window areas.

Other coefficients used for the modeling of walls and ceiling are given in Table 5.6. An adjustment factor for attics was considered for the ceiling due to the presence of inter-floor ceilings and ductwork. The construction of the ceiling was adopted as a 200 mm thick concrete layer. The color of the walls and ceiling was considered light, for which the color adjustment factor K is 1 [20].

For the calculation of the thermal load associated with glazing, the solar heat gain factors (SHGF) from Table 5.7 were used, given as a function of São Paulo's latitude and the orientation of the faces. The window shading coefficient (SC_{window})

Table 5.5 Total wall area and correction factor for latitude for each face of the building. *Source* [3, 4]

Direction	Walls area [m^2]	*LM*
North face	25.28	−3.3
South face	25.28	1.6
East face	21.19	0
West face	42.41	0

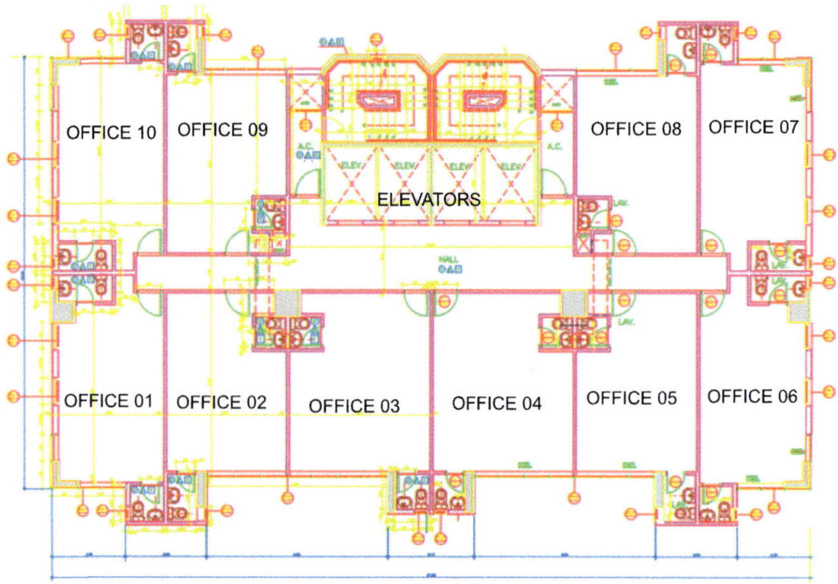

Fig. 5.6 Floor plant of the commercial units distribution in the Philadelphia Building

Table 5.6 Coefficients that were adopted for the walls and ceiling of the modeled building. *Source* [3, 4]

Coefficients	Value
Global heat transfer coefficient of the walls (U_{wall})	3.69 W.m^{-2}.°C
Color factor for the surface of the wall (K_{wall})	1
Global heat transfer of the roof (U_{roof})	0.715 W.m^{-2}.°C
Surface factor of color of the roof (K_{roof})	1
Ceiling attic correction factor (F_{roof})	0.875
Ceiling geographic correction factor (LM_{roof})	0.5
Roof area (A_{roof})	582.18 m^2

Table 5.7 Areas, correction factor (LM), and solar heat gain factor (SHGF) for the windows facing each direction. *Source* [3, 4]

Orientation	Windows area [m^2]	LM	SHGF
North	24.30	−3.3	136
South	39.16	1.6	154
East	50.54	0	669
West	29.95	0	669

Table 5.8 Coefficients used in modeling the building's equipment and lighting. *Source* [3, 4]

Coefficients	Adopted value
Installed lamp power (W_{lights})	5 $W.m^{-2}$
Use factor (F_{ul})	1
Application factor (F_{sa})	1.2
Illumination of the unities ($A_{lights,unit}$)	324 m^2
Equipment density (W_{equip})	5.4 m^2

considered was 0.81. The overall thermal exchange coefficient for windows (U_{window}) adopted was 3.69 Wm^{-2} °C. The term CLF is provided by the literature [20] and varies with the temperature throughout the day. The floors of the modeled building were considered to be of heavy construction in order to obtain the CLF.

The thermal load from equipment and lighting was calculated based on the area of the commercial units (multiplied by 15). Table 5.8 shows the coefficients and parameters adopted in the model.

In order to represent the thermal load from occupants, the heat fluxes associated with the four models of the human body were used (i.e., the sum of the components of radiation, convection, sweat evaporation, and respiration for each individual). The maximum capacity was set as $capac = 730$ people, and the occupancy factor f_{ocup} was defined as a distribution with a peak concentration of people at 12 pm and 1 pm (hours with 100% occupancy), as shown in Fig. 5.7. There is a substantial supposition here; people leave for lunch. Nevertheless, it was outside of the scope of this study to consider the thermal behavior of a person after going to lunch and changing their metabolic behavior (this approach will be employed in future analyses). To obtain the total number of occupants $ocup$, the relationship from $occupants - capability \times f_{occupants}$ was used. The distribution was made as 50% male and 50% female occupants. Among the women, 50% of them were set to be in the luteal phase and 50% in the follicular phase (since, according to [26], the menstrual cycle lasts an average of 28 days and the luteal phase lasts an average of 14 days).

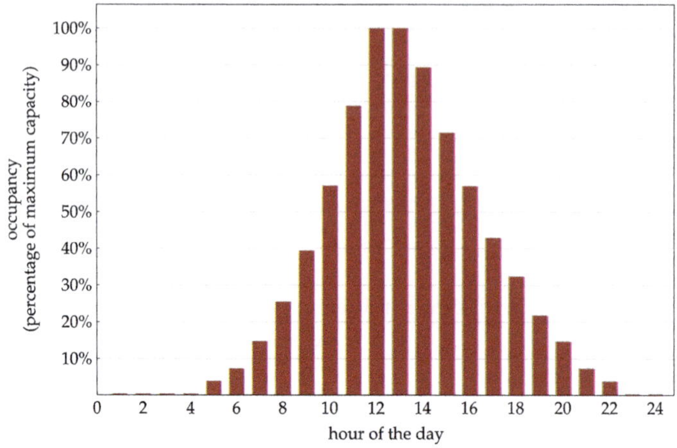

Fig. 5.7 Chart of the building's occupancy profile as a function of the hour of the day. *Source* [3, 4]

5.1.2.1 Results of Thermal Load

Figure 5.8 displays the hourly curves of the thermal loads associated with the roof, walls, glazing, lighting, equipment, and occupants in the building model. This first simulation was obtained by setting the internal temperature to 24 °C. It can be observed that the highest thermal load is associated with the glazing, as the Philadelphia building has a highly glazed architectural facade, which cancels out the intended effect of using bricks (insulating material). The peak thermal load occurs around 5 pm, mainly due to the thermal inertia of the building elements. In addition, the ther-

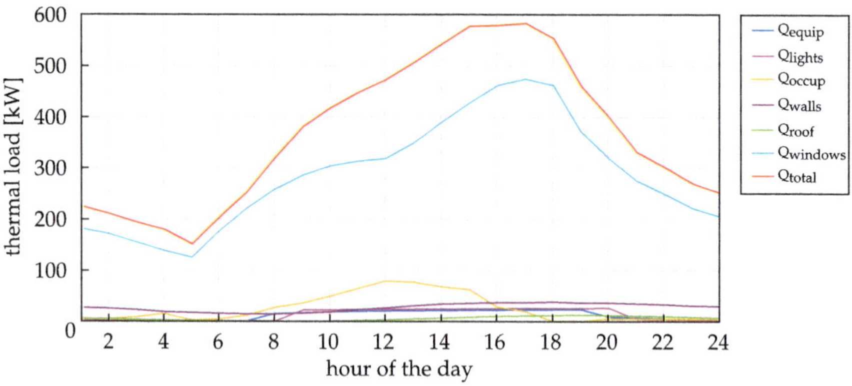

Fig. 5.8 Graph of thermal loads associated with the ceiling, walls, glazing, lighting, equipment, and occupants in the Philadelphia building model and the total thermal load for the operating temperature 24 °C. *Source* [3, 4]

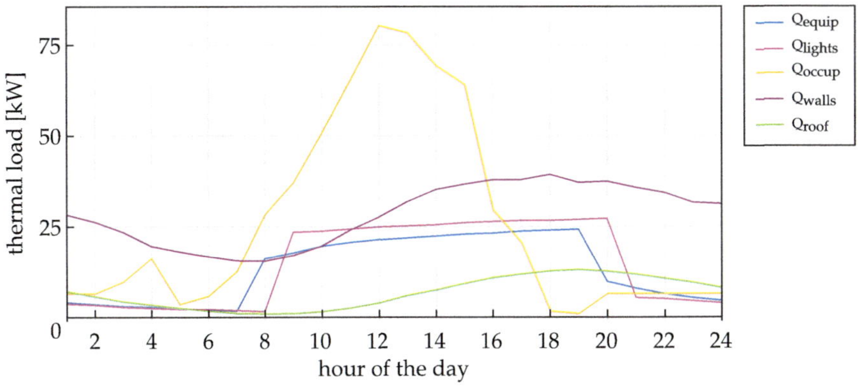

Fig. 5.9 Graph of thermal loads associated with the ceiling, walls, lighting, equipment, and occupants in the Philadelphia building model for the operating temperature of 24 °C. *Source* [3, 4]

mal load from glazing is highest during the illuminated period of the day, due to the incidence of solar radiation. Glazing works act as a greenhouse, being highly transparent to solar radiation and opaque to thermal radiation emitted by occupants, lighting, and equipment, thus retaining energy inside the building.

Since the order of magnitude of the thermal load associated with glazing is greater than that of the others, in order to facilitate the visualization of the behavior of the thermal load curves associated with occupants, lighting, equipment, walls, and ceiling, the corresponding region in the graph in Fig. 5.8 was zoomed in, thus generating Fig. 5.9. It can be observed that the second highest thermal load is associated with the building occupants. The peak occurs between 12 pm and 1 pm, the period of maximum occupancy (bearing in mind that there is a hypothesis of the occupant not leaving the room to avoid additional refinements regarding exercise to go lunch and other aspects), and it is approximately 78 kW. The thermal load curves associated with lights and equipment have a similar behavior, being nearly constant and present only during business hours (i.e., when they are in use). The thermal load curves associated with walls and ceilings have a similar sinusoidal profile, since they are made up of the same mechanisms (conduction, radiation, and convection on the surfaces). It is worth highlighting the "delay" in the peak thermal load caused by the bricks (6 pm) and, therefore, the need to pay attention to such aspects when designing the building. In a tropical country like Brazil and a commercial city such as São Paulo, there must be a balance between the benefits of glazing and the problems with thermal load, mainly in extreme scenarios, as those that occurred in 2023.

5.1.2.2 Exergy Analysis and Adaptation

This use of the models created to represent the occupants of the building implies a more responsive dynamic between the environment and equivalent thermal loads, as for each operative temperature, changes occur not only for the occupants' thermal comfort indicators (such as PMV, PPD, and B_d) but also for the heat fluxes between the body and the environment [3, 4].

For the exergy analysis of the modeled environment, the external environment was taken as the reference environment. In other words, in the expression of the exergy associated with thermal exchanges (walls, ceiling, glazing, lighting, and equipment) given by Eq. (5.12), it was assumed that $T_0 = T_{ext}$. The surface temperature of thermal exchange was assumed to be the internal temperature of the conditioned environment, i.e., the operative temperature T_{op}.

$$\dot{B}_{Q_{walls}} = \dot{Q}_{walls}\left(1 - \frac{T_{ext}}{T_o}\right) \tag{5.12}$$

Following such logic, Molliet and Mady [3, 4] presented Fig. 5.10, which shows the exergy flow rates, where the external environment tends to decrease the exergy of the building, and the air conditioning system supplies exergy to the building in order to keep it at the operative temperature lower than the environmental one. To provide this exergy, the building uses the compression power $\dot{W}_{comp}$ obtained through the conversion of electrical energy.

The vapor compression refrigeration model proposed by [3, 4] consists of the following: a compressor; a condenser, which exchanges heat with a cooling tower; an expansion valve; and an evaporator, which allows thermal exchange between the refrigerant of the cycle and the water that cools the air. The fluid used in the simulations was the refrigerant R22 gas, the same used in the actual refrigeration system installed in the Philadelphia building, composed of two central chillers with a capacity of 100 TR each (i.e., about 351.7 kW each). Therefore, it is possible to draw parallels between the data obtained through simulations and the actual sizing of the equipment.

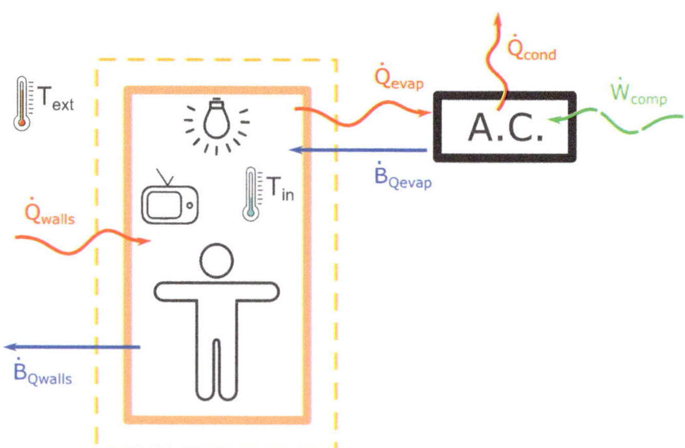

Fig. 5.10 Schematic drawing of the integrated model, considering exergy inputs and outputs. *Source* [3, 4]

5.2 Results and Discussions

As noted in the studies by Martinez et al. [2], Molliet and Mady [3] and others, when comparing the results of the simplified model with the complete model proposed by [1], it is observed that, although the temperatures remain close in the different layers, the profile is constant in the simplified model. This suggests a relatively strong assumption for fat and skin layers, although this temperature represents an average. However, for muscle and core tissues, this simplified approach provides a satisfactory approximation, as discussed in [3] and [2], even though there is a slight temperature variation along the radius. However, it can still be inferred that the overall temperature distribution is similar and adequate for the conditions considered for the simplified model, bearing in mind that these temperatures are a representative average. Another important point is obtaining the energy flow rates to the environment (in this case, the simplification is significantly closer to experimental observations) and internal body temperature for different anatomies. Figure 5.11 shows the exergy rates as a function of operative temperature (T_o) and relative humidity (ϕ), which follows the same trend as validated models [1, 5, 6]. These results also show that for the same amount of muscle, the higher the body mass (fat percentage) will lead to higher destroyed exergy rates.

5.2.1 *Body Composition*

A proposition of two dimension maps, based on the results of [2] are in Figs. 5.12, 5.13 and 5.14, with the estimated mean vote (PMV) and percentage of dissatisfied people (PPD) as a function of the exergy destruction rate $\dot{B}_{dest}$ and exergy transfer to the environment $\dot{B}_{env}$. In these Figures, it is possible to notice that when PMV < 0, regions considered to have a "cold" thermal sensation, relative humidity does not significantly influence the relationship between PMV and $\dot{B}_{dest}$. Furthermore, regions

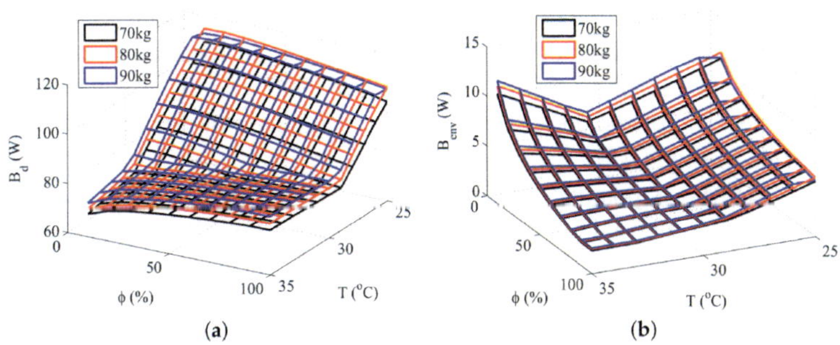

Fig. 5.11 **a** Destroyed exergy rate ($\dot{B}_{dest}$) and **b** exergy transfer to the environment rate ($\dot{B}_{env}$) as a function of operative temperature (T_o) and relative humidity (ϕ). *Source* [2]

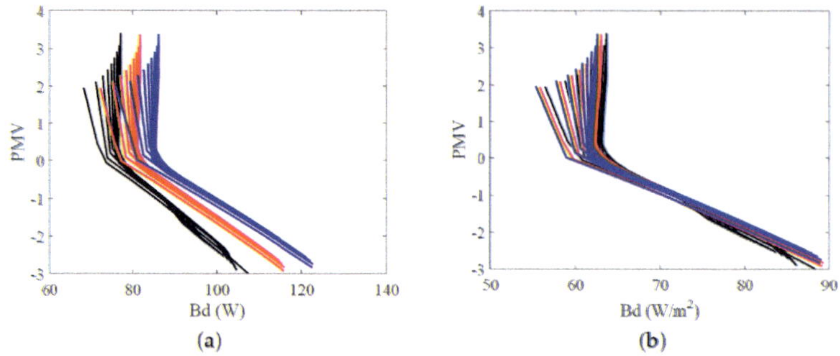

Fig. 5.12 The relation between PMV, with $\dot{B}_env$ for different relative humidities (10–100%) for body masses of 70, 80 and 90 kg, with a constant fat mass of 15 kg: **a** PMV as a function of the destroyed Exergy rate ($\dot{B}_d$ in Watts); **b** PMV as a function of exergy flux rate ($\dot{B}_d/A_d$ in W/m^2). *Source* [2]

with minimum values of $\dot{B}_{dest}$ would coincide with PMV $\to$ 0. The percentage of dissatisfied people is close to the minimum for these points. As PMV values tend towards values greater than zero ("hot" thermal sensation), the rate of exergy destruction and PMV do not have a clear trend. Results of $\dot{B}_{dest}$ close to the minimum (70 to 75 W) would indicate regions that would lead to thermal comfort. However, such points coincide with zones equivalent to the desert, and the PMV is close to magnitudes between 2 and 4. The same trend may be seen in Mady et al. [5]. These results and differences between PMV and $\dot{B}_{dest}$ show that literature results that only destroyed exergy are sufficient to determine thermal comfort conditions are valid for office conditions [22, 30, 31], where we control the relative humidity and temperature. Therefore, another index such as $\dot{B}_{dest}$ can certify thermal comfort in a broader condition, e.g., outdoor conditions, even in thermal stress [5, 42].

Figure 5.13a depicts the PMV as a function of the exergy transfer to environment rate in [W] ($\dot{B}_{dest}$), while Fig. 5.13b shows the PMV as a function of the exergy flux rate in [W/m^2]. Figure 5.13a demonstrates a consistent trend with the concept of higher body mass leading to higher metabolism and higher irreversibilities; nevertheless, the PMV is the same for the three body masses (70, 80, and 90 kg), showing that there must be a modification in this index to be valid for anatomies. Normalizing by surface area leads to similar behavior for PMV and $\dot{B}_{dest}$ independently of the body mass. Therefore, there is a consistency and overlap for both behaviors, as demonstrated in Fig. 5.13b.

When the PMV is positive, it is necessary to use both $\dot{B}_{dest}$ and $\dot{B}_{env}$ to accurately evaluate the thermal environment. At a humidity of 10%, for instance, $\dot{B}_{dest}$ reaches its minimum value for PMV higher than 2, while $\dot{B}_{env}$ has its maximum value for the same PMV range. Contrarily, at a humidity of 100%, $\dot{B}_{env}$ is minimal and PMV higher than 2, although $\dot{B}_{dest}$ is not minimal at those points. Therefore, for the first time in the literature, the idea that the ordinate pair ($\dot{B}_{dest}$, $\dot{B}_{env}$) is a complete tool for assessing the thermal environment. Moreover, there are some points where only the destroyed exergy provides complete information, e.g., office envi-

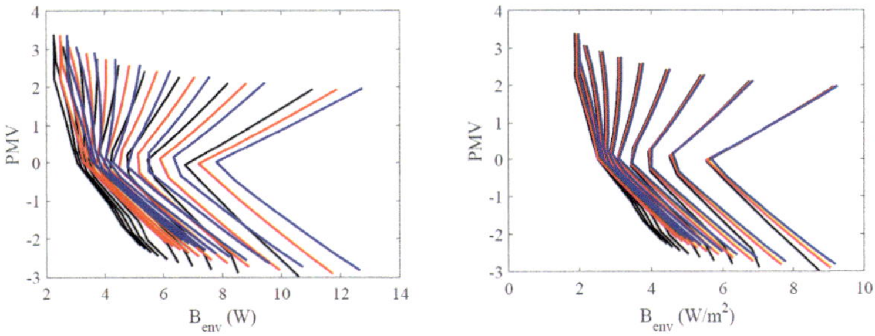

(a) (b)

Fig. 5.13 The relation between PMV, with $\dot{B}_{e}nv$ for different relative humidities (10–100%) for body masses of 70, 80 and 90 kg, with a constant fat mass of 15 kg: **a** PMV as a function of the exergy transfer to environment rate ($\dot{B}_{env}$ in Watts); **b** PMV as a function of exergy flux rate ($\dot{B}_{env}/A_d$ in W/m^2). *Source* [2]

ronments; nevertheless, the comprehensive analysis incorporates the exergy analysis. Additionally, the exergy analysis is more sensitive to relative humidity than the first law of thermodynamics. These analyses using the limits of the model are important for workers, e.g., outside of buildings in stressful climates using Personal Protective Equipment–PPE–as discussed by [18].

Figure 5.14 demonstrates that the response of the model's exergy flow rates in [W], when normalized by surface area [W/m^2], is similar when only one type of tissue (fat or muscle) is modified. Figure 5.14a, b highlight the importance of considering both the destroyed exergy rate ($\dot{B}_{dest}$) and the exergy transfer to environment rate ($\dot{B}_{env}$) to accurately evaluate thermal comfort. It is evident from these figures that $PMV \approx 0$ and $PPD \approx 5\%$ only when both physical quantities are at, or around, their minimum points is it possible to achieve thermal comfort conditions. This condition is especially important for low humidities. For example, when considering $\phi = 10\%$, it can be observed that, when the destroyed exergy is minimal, the exergy transfer rate to the environment remains high, and the person is not in thermal comfort conditions (PMV ≥ 2). Hence, the exergy transfer to the environment highlights the significant role of the thermoregulatory system, which reveals a deviation from thermal comfort even for thermal neutrality, for instance, asymmetric radiation.

One discussion that we must consider is the effect of modifying the muscle or fat mass on the exergy behavior of the body when it is in exergy flux [W/m^2]. The exergy transfer to environment rates is not significantly affected by tissue modification, regardless of the body composition. Nevertheless, the destroyed exergy suffers influences as each tissue has its metabolism. Muscles have higher metabolism per kg, which changes their metabolism for "cold" environments. Therefore, their energy density is higher than that of fat tissues. A clear demonstration in Fig. 5.15a.

The exergy indexes are in Fig. 5.15, to analyze if the area normalization is a suitable generalization of the human body response when subjected to different environments. Figure 5.15a. shows that modifying the body's constitution, especially the muscle mass, affects the destroyed exergy due to increased metabolism, as previously

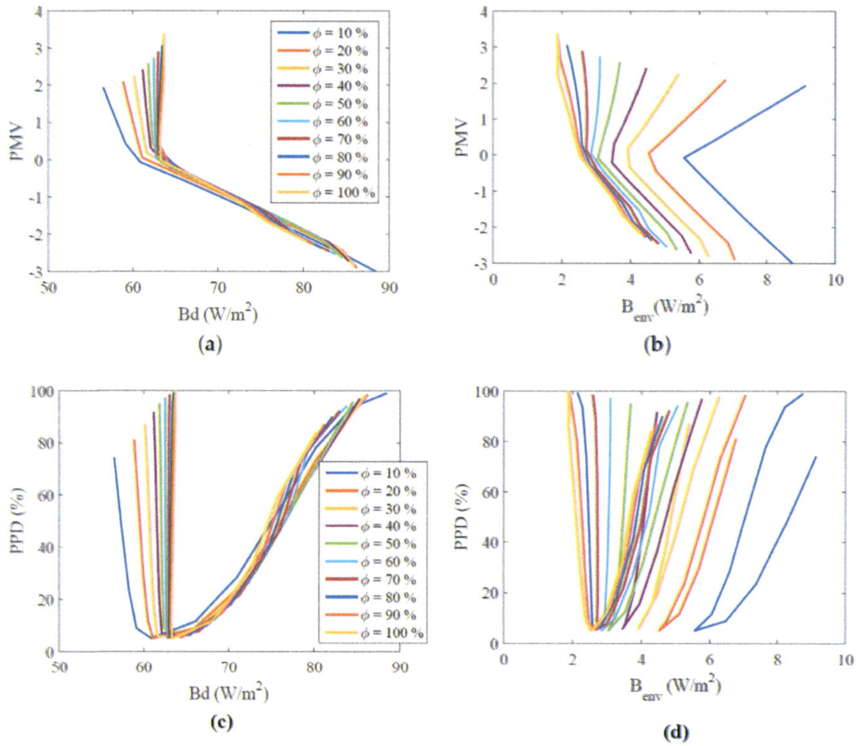

Fig. 5.14 The relation between PMV and PPD, with $\dot{B}_d''$ and $\dot{B}_{env}''$ for different relative humidities (10–100%) for body masses of 70, 80 and 90 kg, with a constant fat mass of 15 kg: **a** PMV as a function of the exergy transfer to environment rate ($\dot{B}_{env}$ in Watts); **b** PMV as a function of exergy flux rate ($\dot{B}_{env}/A_d$ in W/m^2). *Source* [2]

explained. The main effect observed is the translation of the curve, in which the slope is the same; knowing the exergy destroyed at the thermal neutrality condition (PMV = 0), it is possible to obtain a general behavior. Figure 5.15b demonstrates that body composition minimally affects the exergy transfer to the environment when there is a normalization by skin area.

5.2.2 Difference Between Male and Female Responses

In this section, the results obtained from the testing of the developed computational model of the human body's passive and thermoregulatory systems for four types of anatomies and different clothes are presented. Then, these models are tested and compared with the traditional Fanger indices [17]. Finally, the energy expenditure

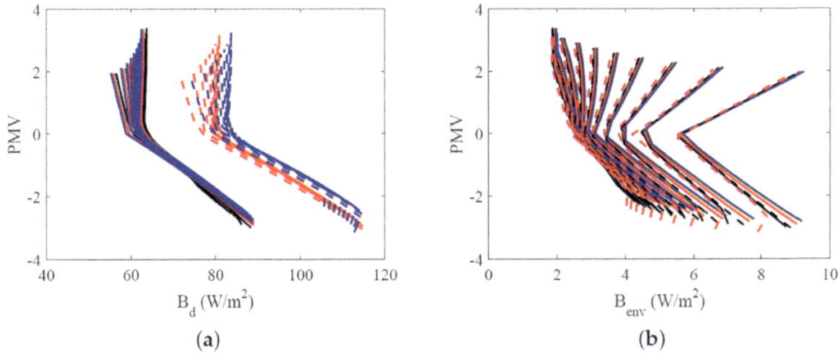

Fig. 5.15 PMV as function of exergy indexes and relative humidity ($10 < \phi < 100\%$) for a body mass of 70, 80 and 90 kg with fat mass of 15 kg (continuous line) and for a mass of 80 and 90 kg and muscle mass of 30 kg (dashed lines): **a** PMV as function of destroyed Exergy; **b** PMV as function of exergy transfer to environment. *Source* [2]

of the building and decrease in the energy expenditure are examined based on the thermal comfort concepts.

Two male models–one dressed in a full suit (MS and 1 CLO) and the other with lighter clothing (ML and 0.67 CLO)–and two female models with the same clothing in the luteal (FL) and follicular (FF) phases were simulated (Fig. 5.16). The comfort (PMV and PPD) and exergy ($\dot{B}_{env}$ and $\dot{B}_{dest}$) indicator curves as a function of the air temperature T_a for the four models are in Fig. 5.16. In all models, the PMV curves increase almost linearly with increases in T_a because this index represents the average rating of the thermal environment. The minimum points on the PPD curves coincide with PMV= 0. An increase/decrease in T_a made several people experience thermal discomfort, as indicated by the PMV and PPD curves. Regarding the exergy behavior, at low T_a values, both $\dot{B}_{dest}$ and $\dot{B}_{env}$ increase, while these values decrease sharply as T_a approaches thermal neutrality, similar to the results obtained by [2]. The shivering increases muscle metabolism at low temperatures, while at temperatures higher than neutrality, metabolism is almost constant. Note that thermal neutrality is considered a thermal comfort condition in this text. Considering the typical temperature and relative humidity ranges of air-conditioned environments ($40\% \leq \phi \leq 60\%$ and 20° C $\leq T_a \leq$ 30° C), the minimum $\dot{B}_{dest}$ values in these Figures can be noted as thermal comfort points. It should be noted that the behavior of $\dot{B}_{env}$ was not conventional and tended to decrease, which was not observed in [2].

This phenomenon is most likely due to the additional thermal resistances considered in this study, i.e., clothing and increased metabolic activities. The point simulated also conducted in relative humidities that the destroyed exergy would be sufficient to determine thermal comfort conditions, which is visible with a slight increase in this parameter for temperatures higher than thermal neutrality.

By comparing the comfortable conditions of FF (Fig. 5.16a) and MS (Fig. 5.16b), the $\dot{B}_{dest}$ and $\dot{B}_{env}$ are higher for MS. Moreover, given that the body mass percentage of muscles is higher for men than women, this difference may be related to metabolic activity. Another difference between these male and female models is in the mini-

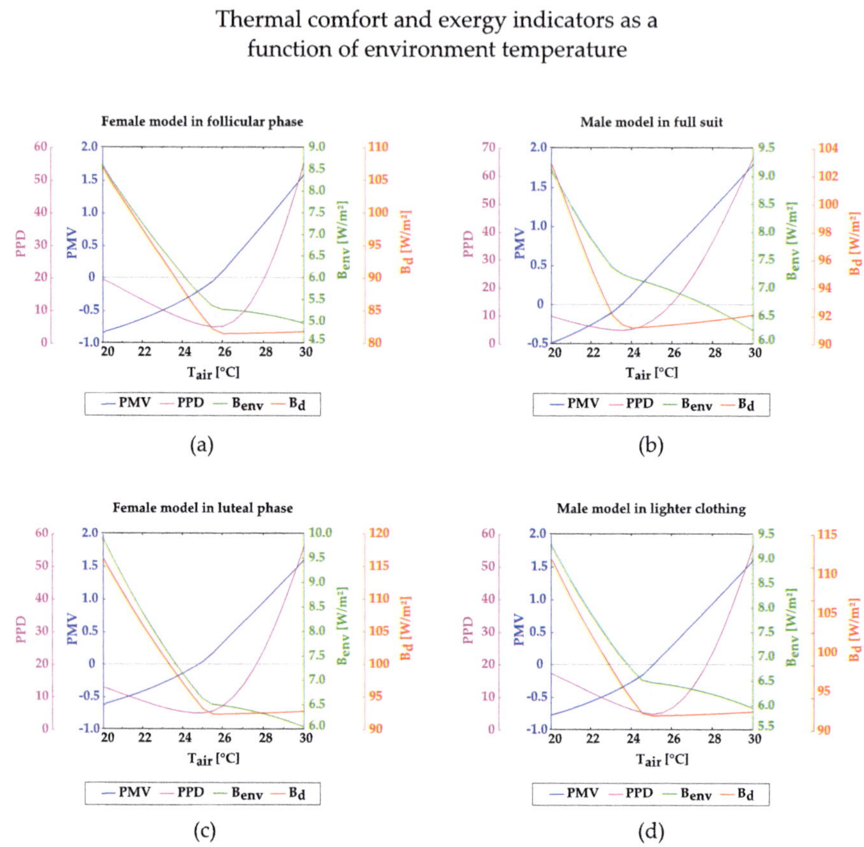

Fig. 5.16 Thermal comfort (PMV and PPD) and exergy ($\dot{B}_{env}$ and $\dot{B}_{dest}$, normalized by body surface area) indicators as functions of the air temperature T_{air} for the models representing: **a** females in the follicular phase; **b** males in a full suit; **c** females in the luteal phase; and **d** males in lighter clothing. *Source* [3]

mum value of $\dot{B}_{dest}$, which indicates the environmental temperatures that result in thermal comfort. Table 5.9 shows all comfort temperatures considering the thermal comfort indicator curves (T_{Fanger}) and exergy curves (T_{exergy}) for all four models. As expected, T_{Fanger} and T_{exergy} are lower for MS than for FF.

The results in Table 5.9 indicate that the differing thermal behaviors of different genders are not clothing-related. When the same clothing for both sexes is used ($I = 0.67$ CLO), FF still requires higher T_{Fanger} and T_{exergy} values to achieve thermal comfort in comparison with ML. This conclusion derived from the exergy analysis was only immediately obvious after this analysis. The difference in males' and females' thermal environment behaviors is related to clothing differences and anatomic conditions. However, lighter clothes may be used for females to make their comfort temperatures similar.

Table 5.9 Thermal comfort temperatures obtained for all four models. *Source* [3]

Model	T_{Fanger} (° C)	T_{exergy} (° C)
FF	25.5	26.0
FL	24.8	25.5
MS	23.5	24.0
ML	25.0	25.0

Nevertheless, it would involve modifying the dress code with personal adjustments to guarantee personal thermal comfort. An example of this modification occurred in Japan with the so-called Cool Biz Program (2005), where the air condition temperature was adjusted to higher values, with a flexibilization in the dressing code. This initiative resulted in a significant reduction in energy consumption and CO_2 emissions.

For the same relative humidity conditions, women experience thermal comfort at lower temperatures in the luteal phase than in the follicular phase, as seen in Table 5.9. During the luteal phase, the increase in blood progesterone concentration causes an increase in metabolic activity, implying higher oxidation rates that generate more irreversible phenomena and, consequently, higher values of $\dot{B}_{dest}$ and $\dot{B}_{env}$. Considering the curves shown in Fig. 5.16a, c, it may be noted that the values of $\dot{B}_{dest}$ near the thermal comfort regions for FL are higher than those for FF.

5.2.3 Direct Application to the Commercial Sector

Figure 5.17 shows a curve of the energy consumption resulting from artificial acclimatization for the Philadelphia Building, whereas simulated herein with CLTD (cooling load temperature difference) as in [32]. The T_{Fanger} and T_{exergy} for the four models are also available in Fig. 5.16. It is also possible to conduct an exergy analysis and compare our models' PMV and PPD values. For both criteria, MS achieved thermal comfort at lower temperatures than the other models, while FF required higher temperatures to experience thermal comfort.

In general, FF requires high temperatures to achieve thermal comfort. The FL and ML models showed similar thermal comfort conditions. This fact indicates that a proper study of thermal comfort conditions may lead to a higher satisfaction rate with the thermal environments in which diversity is increasing. A possible explanation for the observed trends is that even though females have a slower metabolism than males, the increased metabolic activity in the luteal phase causes females to experience comfort in similar environmental conditions.

Regarding the energy consumption in Fig. 5.17, it is possible to note that the energy expenditure decreases linearly as the operating temperature of the air conditioning system increases, which indicates that adopting temperatures close to the comfort conditions of ML and FL (approximately 25.2 °C) would provide a 22.2% lower energy consumption compared to the temperature required for MS (approximately

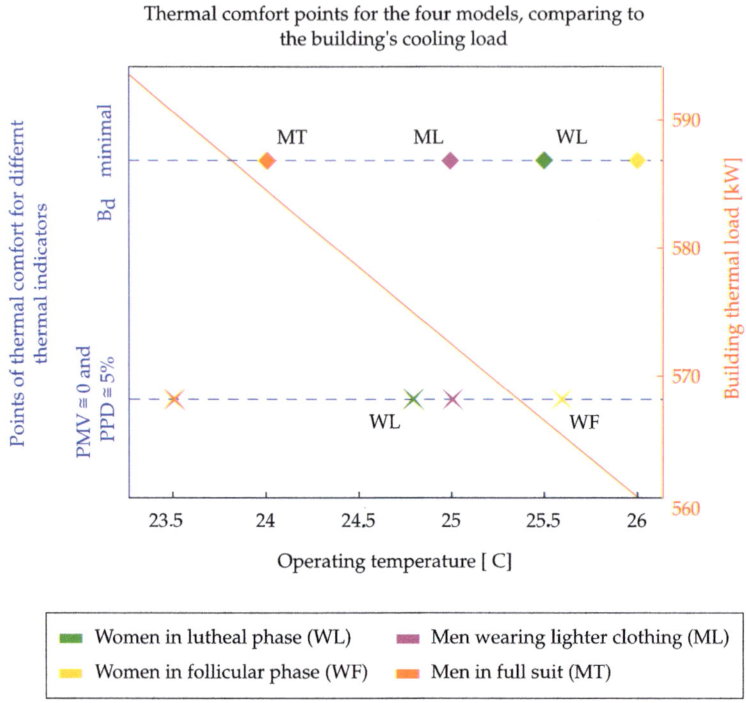

Fig. 5.17 Thermal comfort points for the four models compared to the thermal cooling load curve of the modeled building. This figure is an evolution of the one published by [3], with new data and complete environmental input

23.7 °C). How to achieve this? The first proper idea is to adopt a dress code in which the person would feel more comfortable in this temperature specified. How to achieve that? Get inside the companies, change cultures, avoid any sectarianism, teach the necessity of minimal user control, and several other aspects that only information such as [33] would achieve, that the conditioning system is operating with a capability with lower energy costs, and higher thermal comfort. Why conduct such studies? The same male subject, with a modification of clothes, would permit an increase in the environment temperature from around 23/24 °C to 25 °C. The main result is reducing the carbon footprint associated with the electrical mix to achieve a more comprehensive thermal environment with a rational use of exergy (or the quality of the energy). Differences between male and female responses suffer reduction with the correct freedom of choice; there is always an answer. In that environment, "what clothes will one wear in thermal comfort and respect the dress code of the environment?"

Exergy analysis and its application to the human body is a tool that is already consolidated to evaluate thermal comfort conditions. Studies date from around three decades ago in three main groups, as indicated in [6, 21, 34–36]. Nowadays, several investigations compare exergy analysis with thermal comfort conditions; for instance, differences due to muscle and fat mass composition [2] between males and females can be considered to obtain a more inclusive work environment [3]. Other authors have studied the relationship between CO_2 composition and human body exergy consumption [37], even through the application of CFD analyses to help assess the thermal environment [38]. Some review analyses compared the models in the literature, concluding that a unified exergy method is important to assessing thermal comfort conditions and the building environment using the same indicator [6, 39, 40]. The idea is that minimal points of destroyed exergy rate and exergy transfer to the environment represent thermal comfort conditions [6], and these thermodynamic figures are more sensitive to relative humidity and extreme conditions [6, 42] than the traditional predicted mean vote (PMV) and percentage of dissatisfied people with the actual thermal environment (PPD). Furthermore, minimizing the entropy generation and exergy destruction of the human body may contribute to a better quality of life and lifespan [41].

McCartney and Humphreys [43] point out that for non-repetitive activities, such as those in most offices, it is challenging to define methodologies for measuring productivity without using subjective and/or biased criteria. However, even though it is not possible to accurately assess the degree of increased productivity, it was observed that productivity did vary according to the occupants' perception of temperature. Thus, the literature strongly suggests a link between promoting thermal comfort and productivity, even though establishing these relationships directly and measurably requires a more thorough investigation. In light of this, conducting thermal comfort studies for the adaptation of work environments can not only lead to economic advantages associated with energy savings but also to higher work performance from employees [44].

Thus, it is possible to propose better operating conditions to save energy and provide comfort to more occupants. One way to achieve such an effect more efficiently is to propose changes in clothing habits in conjunction with an increase in the operating temperature of the air conditioning system. By evaluating the same parameters for the human body and the building, but considering that instead of wearing a suit, men are wearing lighter clothes, we have the data in Table 5.10.

Table 5.10 Building thermal load ($\dot{Q}_{total}$), exergy destruction rate ($\dot{B}_d$), estimated average vote (PMV), and percentage of dissatisfied people (PPD) obtained for the building simulation for operating temperatures of 21, 22, 23, 23.5, 24, 25, 25.5, and 26 °C, considering that men are wearing lighter clothes. *Source* [2]

T [°C]	21	22	23	24	25	25.5	26
$\dot{Q}_{total}$ [kW]	639	626	613	599	586	580	574
$\dot{B}_d$ [kW]	588	576	566	536	540	534	529
PMV [−]	−0.60	−0.49	−0.36	−0.21	0.003	0.14	0.28
PPD [%]	12.8	10.2	7.8	5.9	5.0	5.4	6.7

Table 5.11 Clothing insulation values $I_{clothing}$ that lead to comfortable conditions for men and women (in the luteal and follicular phase) for operating temperatures of 21, 22, 23, 24, 25, 25.5, and 26 °C. *Source* [2]

Temperature [°C]	21	22	23	24	25	25.5	26
Men (CLO)	1.35	1.21	1.03	0.84	0.68	0.56	0.49
Women (lutheal) [CLO]	1.36	1.22	1.04	0.85	0.69	0.57	0.50
Women (follicular) [CLO]	1.40	1.23	1.16	0.97	0.78	0.69	0.60

From Table 5.10, it can be observed that, at 25 °C, only 5.01% of the occupants would be dissatisfied with the thermal environment. Now, by fixing the temperatures and varying the clothing insulation I_{roupa} until thermal comfort conditions were found for each occupant model, it is possible to obtain Table 5.11. According to the data in the table, for the operative temperature of 25.5 °C, the ideal clothing is similar for the 4 models. In other words, it is possible to achieve considerable energy savings and, consequently, a more efficient and rational use of energy resources by proposing an adjustment of clothing to the climate and then increasing the operative temperatures of air conditioning.

Table 5.12 indicates the values of the total thermal load and exergy destruction rate associated with heat and mass transfers with finite temperature and concentration differences. It is worth noting that such behavior is to be expected, as the higher the air temperature, the lower the electricity costs. However, the weighted PMV suggests that there is a minimum, which would characterize the optimum point between the exergy destruction rate of each individual and that of the building. It is possible to notice that, from 25 °C onward, there is an increase in the PPD (percentage of dissatisfied people) due to the weighting between male and female occupants. In other words, women would be in conditions closer to comfort at temperatures around 26 °C, while men with light clothing and women in the luteal phase would be more comfortable at a temperature slightly below 25 °C.

Table 5.12 Maximum thermal load values $\dot{Q}_{total}$, power for the compressor $\dot{W}_{comp}$, exergy destroyed in the building $\dot{B}_{d,edif}$, exergy destroyed in the refrigeration cycle $\dot{B}_{d,cycle}$, and total exergy destroyed $\dot{B}_{d,total}$ for operating temperatures of 21, 22, 23, 24, 25, and 26 °C. *Source* [2]

	21 °C	22 °C	23 °C	24 °C	25 °C	26 °C
$\dot{Q}_{total}$ [kW]	638	624	611.27	598.60	586.21	574.01
$\dot{B}_{d,build}$ [kW]	587	575	565	551	540	529
$\dot{B}_{d,cycle}$ [kW]	62	61	60	58	57	56
$\dot{B}_{d,total}$ [kW]	649	636	716	609	597	585
$\dot{W}_{comp}$ [kW]	168	165	161	158	155	151.5

5.3 Concluding Remarks

Several authors have been performing exergy analysis on the human body to obtain additional indicators of thermal comfort, complementing those proposed in the energy literature, such as PMV and PPD. Correlations based on minimum exergy destroyed have shown some adherence to thermal comfort conditions. In this study, we conducted some comparisons among anatomies (fat and muscle percentage) and between male and female residents with different wearings, most notably.

Therefore, a simplification of the model by Ferreira and Yanagihara [1] proposed in [2] and [3] to evaluate thermal comfort conditions. Therefore, it is possible to conclude that:

- The effects of body composition were successfully related to the parameters of exergy analysis. It was observed that the body's metabolism influences the destruction of exergy and the rate of exergy transfer to the environment, increasing both terms. This relationship can be observed when comparing that an increase in muscle mass causes a greater increment in the indicators based on the Second Law of Thermodynamics than the same increase in fat mass.
- The results obtained in this chapter demonstrate that normalizing by the skin surface area is a good generalization of the human body's response when subjected to different environments. Therefore, exergy analysis and its relationship with the PMV and PPD indices can be used as a general tool to assess thermal comfort conditions. Caution should be taken with the exergy destruction rate when there is a high percentage of muscle mass, as it may cause the surface area to not allow thermodynamic behavior to be independent of body composition.
- Points of minimum exergy destruction do not always occur for thermal comfort conditions. It was found that, when the relative humidity is above 70%, the exergy destruction is minimum for PMV = 0; however, when the humidity is below 70% the exergy destruction is close to its minimum value, but the exergy transfer to the environment is not minimum. Therefore, both terms need to be analyzed for an assessment of thermal comfort conditions.
- For environments considered "cold" (PMV $\leq$0), it was shown that the relationship between the exergy destruction rate and the estimated mean vote is independent of relative humidity. For such environments (temperatures below thermal neutrality), the PMV is zero (or close to zero) when Bd is at a minimum.
- For the so-called "hot"" environments (temperatures above thermal neutrality, i.e., PMV $\geq$ 0), it is necessary to analyze both the exergy destruction rate and the transfer of exergy to the environment in order to have an accurate assessment of thermal comfort. This demonstrates that the exergetic analysis is more dependent on environmental conditions than the terms of the First Law of Thermodynamics.
- The thermal environment in which men achieve thermal comfort has a lower temperature than that required for women. However, women in the luteal phase require lower temperatures than women in the follicular phase to achieve thermal comfort.

- Encouraging the adaptation of a dress code for lighter garments in corporate environments will have a positive impact on energy consumption considering the increase in the temperature of the thermal environment.
- Increasing the environmental temperature can reduce energy consumption and promote more inclusive environments wherein occupants of both sexes are near their thermal comfort conditions.

References

1. Ferreira MS, Yanagihara JI (2009) A transient three-dimensional heat transfer model of the human body. Int Commun Heat Mass Transf 36(7):718–724
2. Martinez-Garcia M, Une RY, Oliveira-Junior S, Mady CEK (2018) Exergy analysis and human body thermal comfort conditions: Evaluation of different body compositions. Entropy 20:265
3. Molliet DS, Mady CEK (2021) Exergy analysis of the human body to assess thermal comfort conditions: Comparison of the thermal responses of males and females. Case Stud Thermal Eng 25:100972
4. Molliet DS, Mady CEK (2024) Investigation of a Building with Male and Female Residents to Achieve Equality of Thermal Sensation Associated with Rational Use of Exergy and Energy. Buildings 14(4):1149
5. Mady CEK, Albuquerque C, Fernandes TL, Hernandez AJ, Yanagihara JI, Oliveira-Junior S (2014) Human body exergy analysis and the assessment of thermal comfort conditions. Int J Heat Mass Transf 77:577–584
6. Ribeiro TJS, Mady CEK (2022) Comparison among exergy analysis methods applied to a human body thermal model. Energy 239:122446
7. Pennes HH (1948) Analysis of tissue and arterial blood temperatures in the resting human forearm. J Appl Physiol 1(2):93–122
8. Chen MM, Holmes KR (1980) Microvascular contributions in tissue heat transfer. Ann New York Acad Sci 335(1):137–150
9. Wissler EH (1985) Mathematical simulation of human thermal behavior using whole body models. Heat Transf Med Biol 1(13):325–373
10. Werner J, Buse M (1988) Temperature profiles with respect to inhomogeneity and geometry of the human body. J Appl Physiol 65(3):1110–1118
11. Gagge AP, Stolwijk JA, Nishi Y (1972) An effective temperature scale based on a simple model of human physiological regulatiry response. Memoirs of the Faculty of Engineering, Hokkaido University 13:21–36
12. DuBois D (1916) A formula to estimate the approximate surface area if height and weight be known. Arch Intern Med 17:863–871
13. Nadler SB, Hidalgo JU, Bloch T (1962) Prediction of blood volume in normal human adults. Surgery 51(2):224–232
14. Kyle UG, Genton L, Hans D, Karsegard L, Slosman DO, Pichard C (2001) Age-related differences in fat-free mass, skeletal muscle, body cell mass and fat mass between 18 and 94 years. European J Clin Nutrition 55(8):663–672
15. Janssen I, Heymsfield SB, Wang Z, Ross R (2000) Skeletal muscle mass and distribution in 468 men and women aged 18–88 yr. J Appl Physiol 89:81–88
16. Flegal KM et al (2009) Comparisons of percentage body fat, body mass index, waist circumference, and waist-stature ratio in adults. American J Clin Nutrition 89(2):500–508
17. Fanger PO (1970) Thermal comfort. Analysis and applications in environmental engineering, McGraw-Hill, New York

18. Yin Y, Zhang D, Zhen M, Jing W, Luo W, Feng W (2022) Combined effects of the thermal-acoustic environment on subjective evaluations in outdoor public spaces. Sustain Cities Soc 77:103522
19. Spitler JD, McQuiston FC, Lindsey KL (1993) CLTD/SCL/CLF cooling load calculation method. ASHRAE Trans 99:183–192
20. American Society of Heating, Refrigerating, Air-Conditioning Engineers (2005) ASHRAE handbook
21. Simone A, Kolarik J, Iwamatsu T, Asada H, Dovjak M, Schellen L, Shukuya M (2011) Olesen B W A relation between calculated human body exergy consumption rate and subjectively assessed thermal sensation. Energy Build 43:1–9. https://doi.org/10.1016/j.enbuild.2010.08.007
22. Shukuya M (2012) Exergy: theory and applications in the built environment. Springer Science & Business Media
23. Indraganti M, Humphreys MA (2021) A comparative study of gender differences in thermal comfort and environmental satisfaction in air-conditioned offices in Qatar, India, and Japan. Build Environ 206:108297
24. Çeter AE, Özbey MF, Turhan C (2023) Gender inequity in thermal sensation based on emotional intensity for participants in a warm mediterranean climate zone. Int J Therm Sci 185:108089
25. Hema Latha NR (2006) Effects of different phases of menstrual cycle on basal metabolic rate. PhD Thesis, Rajiv Gandhi University of Health Sciences
26. Slauterbeck JR, Fuzie SF, Smith MP, Clark RJ, Xu KT, Starch DW, Hardy DM (2002) The menstrual cycle, sex hormones, and anterior cruciate ligament injury. J Athletic Train 37(3):275
27. Frascarolo P, Schutz Y, Jequier E (1990) Decreased thermal conductance during the luteal phase of the menstrual cycle in women. J Appl Physiol 69(6):2029–2033
28. Solomon SJ, Kurzer MS, Calloway DH (1982) Menstrual cycle and basal metabolic rate in women. American J Clin Nutrition 36(4):611–616
29. Bisdee JT, Garlick PJ, James WPT (1989) Metabolic changes during the menstrual cycle. British J Nutrition 61(3):641–650
30. Prek M, Butala V (2017) Comparison between Fanger's thermal comfort model and human exergy loss. Energy 138:228–237
31. Prek M (2006) Thermodynamical analysis of human thermal comfort. Energy 31(5):732–743
32. Ahmed T (2012) Software development for cooling load estimation by CLTD method. IOSR J Mech Civil Eng 3(6):1–6
33. ENCE–Etiqueta Nacional de Conservação de Energia (2013) Selo PROCEL de Economia de Energia / ANEXO DA PORTARIA INMETRO No 410/2013, http://www.inmetro.gov.br/legislacao/
34. Prek M (2004) Exergy analysis of thermal comfort. Int J Exergy 1:303–315. https://doi.org/10.1504/IJEX.2004.005559
35. Prek M, Butala V (2017) Comparison between Fanger's thermal comfort model and human exergy loss. Energy 138:228–237
36. Shukuya M (2009) Exergy concept and its application to the built environment. Build Environ 44:1545–1550. https://doi.org/10.1016/j.buildenv.2008.06.019
37. Bajc T, Kerčov A, Gojak M, Todorović M, Pivac N, Nižetić S (2023) A novel method for calculation of the CO2 concentration impact on correlation between thermal comfort and human body exergy consumption. Energy Build 294:113234
38. Xu C, Xie Y, Huang S, Zhou S, Zhang W, Song Y, Luo Y, Tian Z (2023) A coupled analysis on human thermal comfort and the indoor non-uniform thermal environment through human exergy and CFD model. J Build Eng 74:106845
39. Guo H, Luo Y, Meggers F, Simonetti M (2019) Human body exergy consumption models' evaluation and their sensitivities towards different environmental conditions. Energy 183:1075–1088
40. Dutta A, Chattopadhyay H (2021) A brief on biological thermodynamics for human physiology. J Biomech Eng 143:070802

41. Deyranlou A, Revell A, Keshmiri A (2023) Exergy destruction in atrial fibrillation and a new 'Exergy Age Index'. J Theor Biol 575:111623
42. Jia S, Wang Y, Wong NH, Weng Q (2025) A hybrid framework for assessing outdoor thermal comfort in large-scale urban environments. Landscape Urban Plan 256:105281
43. McCartney KJ, Humphreys MA (2002) Thermal comfort and productivity. Proc Indoor Air 2002:822–827
44. Wu X, Zhao J, Olesen BW, Fang L (2013) A novel human body exergy consumption formula to determine indoor thermal conditions for optimal human performance in office buildings. Energy Build 56:48–55

Chapter 6
Circulatory System

6.1 Introduction

The first article of the group regarding the exergy behavior of the heart was proposed by [2]. Later, Roll et al. [1] presented a method to assess the exergy destruction in the heart due to aortic valve stenosis, where stenosis is a reduction in elasticity or stiffening of the heart valve. Thus, there is an increase in pressure drop throughout the circulatory system and its complexities. For such a purpose, the study adapted the model proposed in Fig. 3.1 to that shown in Fig. 6.1.

Based on [2], we propose a process diagram representing the heart to obtain a pressure-volume cardiac cycle diagram for healthy patients and those with some pathology. Figure 6.2 demonstrates the heart divided into four chambers: left ventricle, right ventricle, left atrium, and right atrium. Therefore, changes in blood hemodynamics and valves cause changes in the second law of thermodynamics that are not observed only by the first law's application. The valves are isenthalpic. These heart divisions are traditionally referred to as the right and left heart in the medical literature, as seen in [3]. Still, regarding Fig. 6.2, systemic circulation refers to circulation throughout the body (as the heart delivers oxygen-rich, arterial blood to the entire body). Pulmonary circulation refers to the transport of venous blood to the lungs as the heart pumps oxygen-poor blood to undergo gas exchange, turning venous blood into arterial blood. This chapter adds to the central reservoir of Fig. 5.2.

Figure 6.3 illustrates a schematic representation of a cardiac cycle of the left ventricle, adapted from [4]. Note that the external work or power can be obtained from the area between the dark lines in Fig. 6.3, that is, the area enclosed in diagram $P\,x\,\forall$. The cardiac cycle is composed of four variations of states, both for the right and left sides. It starts with the filling period, followed by an increase in isovolumetric pressure until the internal pressure becomes higher than the circulatory pressure, initiating the ejection period. Finally, there is the isovolumetric relaxation. Such

C. E. Keutenedjian Mady, *Exergy Analysis*, Green Energy and Technology,
https://doi.org/10.1007/978-3-032-01646-1_6

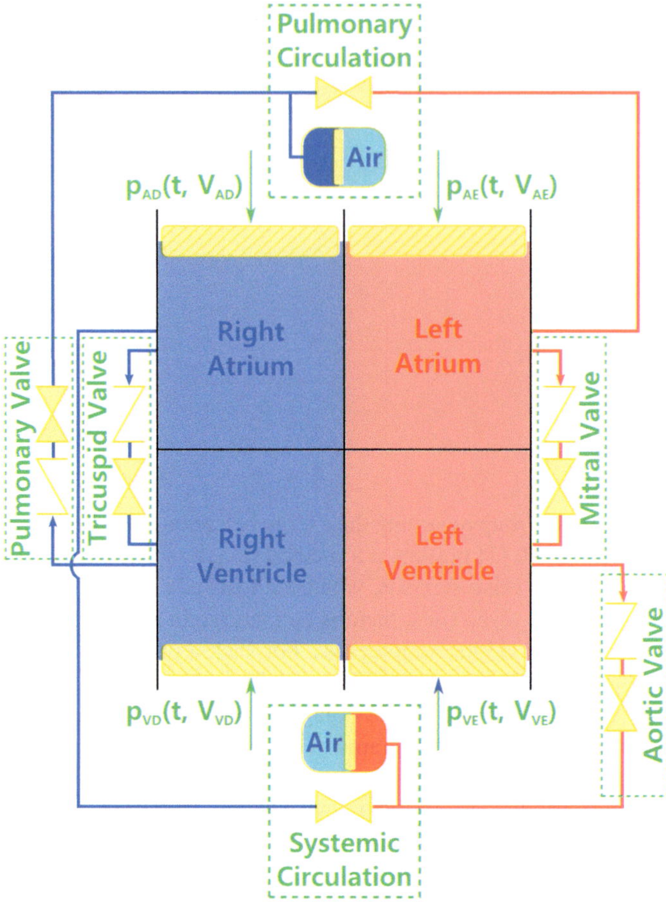

Fig. 6.1 Schematization of the cardiovascular system and intra and intersystemic exergy flows. *Source* [1]

cycle is indicated by the shaded area, and its numerical value represents the work performed by the heart after an external electrical impulse.

From the pressure-volume diagram, the work performed in the cardiac cycle can be obtained. For this text, we considered that the work is from the heart to the blood activated through a passive mechanism triggered by an electrical signal entering the control volume. Equation 6.1 represents this term, noting that both pressure and volume are interdependent and time-dependent. According to [5], cardiac muscle cells contract through electrical signals originating from specialized myocardial cells that function as a kind of controller. The same author also states that the voltage has a gradient in the cell membranes of approximately –80 mV under resting conditions. This value can be modified when ions such as sodium, potassium, and calcium cross the membrane of cardiac muscle cells. The potential difference can reach values

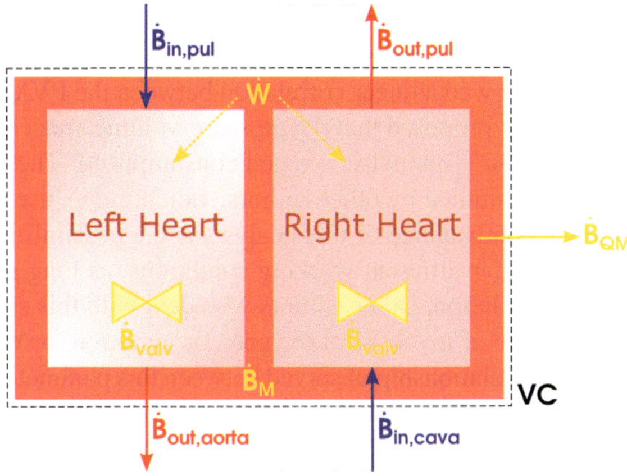

Fig. 6.2 Schematic model of the heart for application of exergy analysis

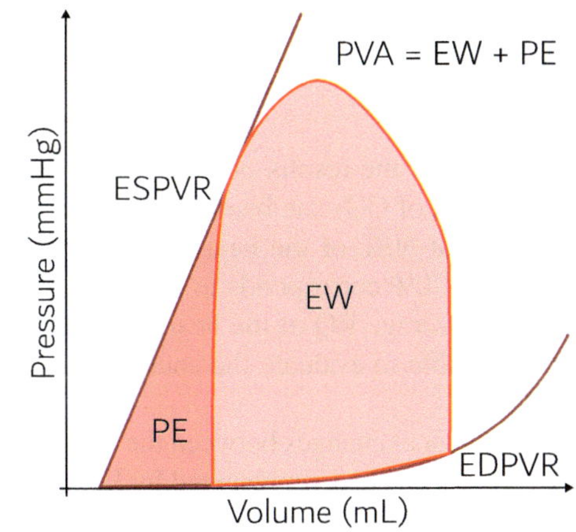

Fig. 6.3 Schematic representation of the cardiac cycle (ventricle pressure-volume diagram). The dark lines represent end-systolic and end-diastolic pressure-volume relationships, while the shaded area represents PVA. *Source* Own author adapted from [4]

close to 30 mV through this process, with an electric current ranging from 0 to 5 mA. Therefore, for the heart to perform some activity, it needs to be a "trigger" to stimulate the muscles, as discussed in detail in [6].

$$\dot{W}_{heart} = \int_{\forall_{min}}^{\forall_{max}} p(\forall) d\forall \tag{6.1}$$

The remaining shaded area in Fig. 6.3 is referred to as potential energy (PE) in medical literature. The dark lines represent the maximum systolic and diastolic pressure of the heart, which are obtained experimentally, as pointed out by [3]. The complete area under the pressure-volume curve is referred to as PVA and is used as a basis for calculating the oxygen consumption of the heart.

$$PVA = \dot{W}_{heart} + E\dot{P} \tag{6.2}$$

Suga et al. [7] analyzed the relationship between PVA and cardiac oxygen consumption. From measurements obtained in a preparation with canine hearts, the authors showed a linear correlation between the PVA area and oxygen consumption. It was then proposed that the pressure-volume area could be used as a potential indicator of $\dot{\forall}_{O_2}$ (volumetric oxygen consumption). The linearity of the correlation has also been studied by other authors, validating its use in studies on cardiac behavior. The great advantage of this analysis is the possibility of extrapolating these data to a heart under different working conditions, as long as it is within the data range of the interpolation. Other authors who deal with this subject include [8–10].

The rate of myocardial oxygen consumption, or $\dot{\forall}_{O_2}$, can be determined through the linear relationship observed between this parameter and the pressure-volume area (PVA) of the left ventricle. In this modeling, the correlation shown in Eq. 6.3, obtained by [9] is used, with a linear correlation coefficient of 0.991. The production of carbon dioxide is estimated based on the consumption of oxygen through respiratory and molar relationships.

$$\dot{\forall}_{O_2} = (1.82 \cdot 10^{-5})PVA + 0.0284 \tag{6.3}$$

Based on the results obtained from Eq. 6.3 and the data presented in [10], the production of CO_2 can be obtained using the respiratory quotient (RQ) of the heart, noting that 30% of the heart's nutrient consumption corresponds to glucose (RQ = 1) and 70% corresponds to lipids (RQ = 0.696). Thus, it was possible to assume that the average RQ of the heart is 0.787. Using the correlations proposed by [11], it is possible to evaluate the energy and exergy metabolism according to Eqs. 3.27 and 3.28.

The heat exchanges between the heart and the human body can be evaluated from the energy balance, represented by Eq. 6.4. The exergy associated with heat transfer can then be found using Eq. 6.5, where T_0 represents the reference temperature (298 K) and T_{heart} represents the average temperature of the human heart.

$$\dot{Q}_{M_{heart}} = \dot{M}_{heart} - \dot{W}_{heart} \tag{6.4}$$

$$\dot{B}_{Q_{M,heart}} = \dot{Q}_{M_{heart}} \left(1 - \frac{T_0}{T_{heat}}\right) \tag{6.5}$$

The variation of exergy associated with the inward and outward flows is divided into a physical component and a kinetic component (Eq. 6.6)–ignoring the variation of potential energy. Blood (subscript s) is assumed to be an incompressible substance. Thus, the physical component can be calculated using Eq. 6.7. The kinetic component, Eq. 6.8, considers the variation of blood exergy due to the difference in velocities between inflow and outflow. The volumetric blood flow rate $\dot{\forall}_s$ [m^3/kg] is determined by the difference between the maximum and minimum volumes of the ventricle, considering the period of one cardiac cycle as the reciprocal of the heart rate. The temperature difference between inflow and outflow is obtained using the

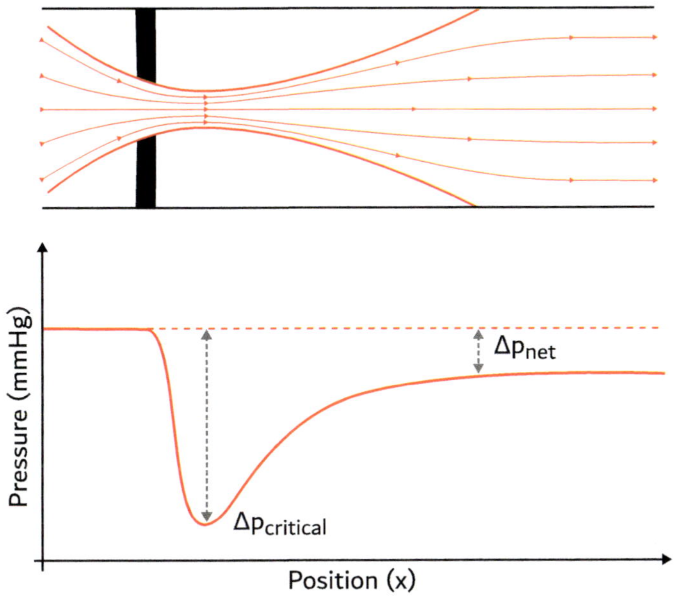

Fig. 6.4 Pressure in a flow passing through an orifice valve concerning position. *Source* [1]

First Law of Thermodynamics, as emphasized in Eq. 6.11. Note that v_s and c_s represent the specific volume [m³/kg] and specific heat of blood [kJ/(kgK)], respectively.

$$\Delta \dot{B}_{bl} = \Delta \dot{B}_{ph} + \Delta \dot{B}_{E_k} \tag{6.6}$$

$$\Delta \dot{B}_{ph} = \left(\frac{\dot{\forall}_{bl}}{v_{bl}}\right)\left[c_{bl}(T_{out} - T_{in}) - T_0 c_{bl} ln\left(\frac{T_{out}}{T_{in}}\right) - \frac{T_0 v_{bl}}{T_{heart}}(P_{out} - P_{in})\right] \tag{6.7}$$

$$\Delta \dot{B}_K = \left(\frac{\dot{\forall}^3_{bl}}{2v_{bl}}\right)\left(\frac{1}{A^2_{out}} - \frac{1}{A^2_{in}}\right) \tag{6.8}$$

The portion corresponding to the destruction of exergy in the heart valves can be estimated using the approach proposed by [12], which approximates the behavior of blood flow through the valves to that of a fluid through an orifice plate, allowing an estimation of the pressure drop during contractions. Figure 6.4 shows the behavior of pressure as a function of flow position, serving as a basis for the formulation. In this figure, it is possible to notice that the $\Delta P_{max} = TPG$ occurs just after the valve, also known as the *vena contracta*, and then stabilizes at a level below the downstream flow. It is important to note that the gradient reaches its minimum at the point where the cross-sectional area of the flow is minimum, also called *vena contracta* or effective orifice area (EOA). Therefore, we have the ΔP_{liq} due to the passage of blood through the valve, which will be used for modeling purposes, since it is a quantity that can be easily obtained experimentally and is available in the medical literature.

Based on the work by [1], we can obtain Eq. 6.9, also known as the "Gorlin Equation". In this equation, AVA represents the area of the valve, in this case, the alveolar valve; C, C_v, and C_c are constants; ρ_{bl} is the specific mass of the blood; $\dot{V}_{bl}$ represents the volumetric flow rate of blood; HR is the heart rate; t_E is the time during which the ventricle or atrium remains in ejection; and ΔP represents the pressure drop across the valve. A constant K (Eq. 6.10) can be used to simplify the equation and was obtained by [12]. This constant assumes a value of 44.3 for the aortic valve and 37.7 for the mitral valve, both in $cm/s\sqrt{mmHg}$.

$$AVA = \frac{1}{CC_vC_C}\sqrt{\frac{\rho_{bl}}{2}}\frac{\dot{\forall}_{bl}}{HRt_E}\frac{1}{\sqrt{\Delta P}} \tag{6.9}$$

$$K\alpha C_v C_c\sqrt{\frac{1}{\rho_{bl}}} \tag{6.10}$$

Regarding the variation in the values of AVA (aortic valve area), [13] showed that, for individuals with a high severity prognosis of aortic stenosis, the range of values found is between 0.24 and 0.93 cm^2, similar to the findings by [14] regarding the development of stenosis-related symptoms, that is, $AVA < 1$. For healthy individuals or those with minor changes associated with aortic stenosis, the values reported in [15] for AVA ranged from 3 to 4 cm^2. These data are consistent with the medical literature [1]. They claim that values ranging from 3 to 4 cm^2 indeed represent healthy individuals. However, it is necessary to consider that small reductions in valve area do not necessarily lead to a sudden decrease in systolic flow and pressure drop to a point of severity where the behavior becomes exponential. Thus, the variation of AVA values (for healthy individuals) could be used as shown in Fig. 6.5, representing the behavior of the valve from a healthy case to a stenotic one. Some authors [16] recommend considering areas larger than 1.2 cm^2 as normal, between 1.2 and 0.8 cm^2 as possible cases of stenosis and smaller than 0.8 cm^2 as indicative of significant stenosis. Further information regarding these differences and how they can be characterized and treated can be found in [17].

The first law of thermodynamics changes from Eq. 2.1 to a simplified form, Eq. 6.11, from which it is possible to obtain the phenomenological behavior of the heart.

$$\rho_{bl}\dot{V}_{bl}c_{bl}\left(T_{out} - T_{in}\right) = \rho_{bl}\dot{V}_{bl}\left[\frac{V_{out}^2 - V_{in}^2}{2} + v_{bl}\left(P_{out} - P_{in}\right)\right] + \dot{M}_{heart} - \dot{W}_{heart} \tag{6.11}$$

Once the pressure drop in the valves is known, the exergy destruction in each of them can be estimated using Eq. 6.12. It is assumed that passage through the valve results in a change in the internal energy of the fluid and consequent heat transfer to the rest of the body, which generates irreversibilities associated with this process. Given the greater difficulty in finding information for the right heart, to determine the exergy destroyed in all valves, a proportionality between the right and left heart is considered. This is based on the relationship between the cardiac work of the right and left heart. The First Law of Thermodynamics can be applied to the heart according to

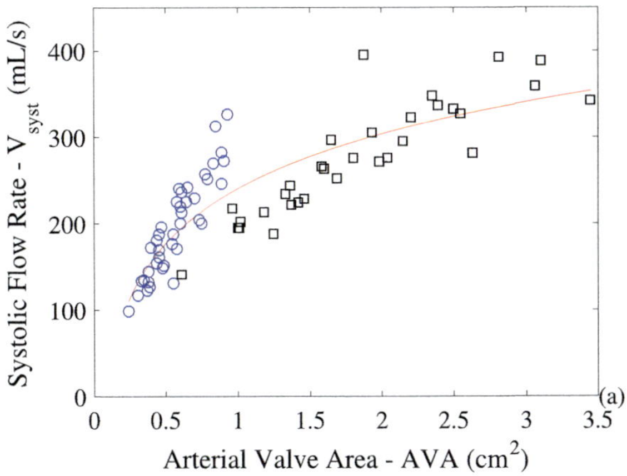

Fig. 6.5 Logarithmic, linear regression between systolic volumetric flow and aortic valve area was measured in 40 patients with severe stenosis (black) and 32 prosthesis operating points (healthy or mildly stenotic patients). R^2 of 0.7497. *Source* [1]

its general form and by using the data available in the medical literature. One can then apply exergy analysis according to Eq. 6.13, and thus evaluate the irreversibilities associated with the flow of blood within the heart.

$$\dot{B}_{d,valv} = \left(\frac{T_0}{T_{int}}\right) \dot{V}_{bl} \Delta P \tag{6.12}$$

For an analysis of the exergy efficiency of the heart, two distinct approaches are considered. Equation 6.14 is suggested by [2] and considers cardiac work as the desired effect, and metabolism as the source of exergy for the myocardium. It is worth noting the similarity to the energy-based efficiency, which is the ratio of work to energy metabolism (taking into account the relationship $M \approx B_M$, we can infer that the two expressions will be linearly dependent).

$$\dot{B}^{heat}_{dest} = \sum \dot{B}_{d,valv} + \dot{B}_{M,heat} + \dot{B}_{Q_{M_{heat}}} + \dot{W}_{heat} - \Delta \dot{B}_{bl} \tag{6.13}$$

Equation 6.15 is proposed by [1], and considers the desired effect as the exergy acquired by the blood during its passage through the heart, or $\Delta \dot{B}_{fluxo}$, divided by the exergy metabolism. In the latter equation, it becomes clear that it is impossible to obtain the increase in capacity (or quality) of work performed by the blood solely using the first law; that is, the efficiency proposed by Eq. 6.15 provides additional information compared to the direct application of the first law.

$$\eta_{b,1} = \frac{\dot{W}}{\dot{B}_M} \tag{6.14}$$

$$\eta_{b,2} = \frac{\Delta \dot{B}_{bl}}{\dot{B}_M}. \tag{6.15}$$

6.2 Results and Discussions

In relation to the effect of a stenosis on the heart, [1] obtained data of volumetric flows as a function of the restriction in area (AVA) for individuals with stenosis, but there were some difficulties in healthy individuals. To do so, the authors found data in the literature for heart valves implanted in patients that would represent the behavior of a healthy heart.

Using the results from [13] a linear regression was proposed for 40 patients with severe stenosis. Now, probably due to the lack of interest in studying the correlation between cardiac output and healthy valve area, another 32 points were obtained from a prosthesis called *Bileaflet* from MEDLINE data obtained from [17]. This type of prosthesis is well-known in the field and has already been validated for different operating points; thus, data from healthy and diseased patients were obtained. The linear regression given by Eq. 6.16 represents the volumetric flow rate (or cardiac output) as a function of the aortic valve area. From Fig. 6.5, it was possible to apply Eq. 6.9 to obtain the pressure drop (or transvalvular pressure gradient–TPG) as a function of the aortic valve area. Figure 6.6 demonstrates these results of pressure drop as a function of the aortic area. Data in circles are calculated, while triangles are prosthesis operating points.

$$\dot{\forall}_{sist} = 91.28 \times ln(PVA) + 241.02 \tag{6.16}$$

In Fig. 6.7a, the same trend of exergy destruction in relation to the severity of the stenosis was obtained as in the results obtained by [1], but the values in this text are higher than those indicated in [1]. This is due to the consideration of not only the flow, but also other parameters of the heart, including the power executed and variable metabolisms with the systolic flow and valve area. It is noteworthy that the power executed by the heart reaches higher values than the destroyed exergy found in

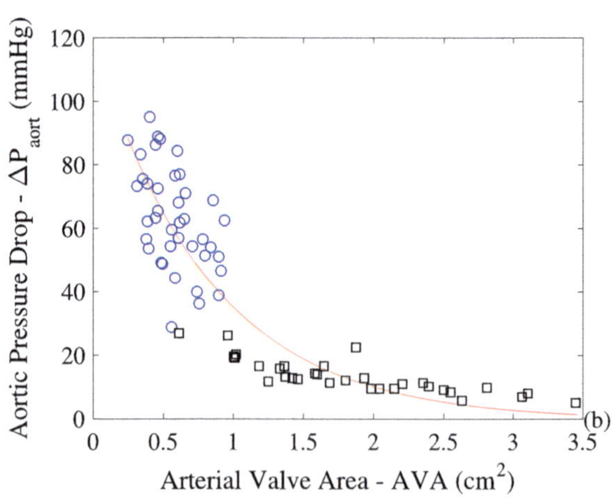

Fig. 6.6 Chart showing the pressure drop as a function of the aortic area. Data in circles are calculated, while triangles are prosthesis operating points. *Source* [1]

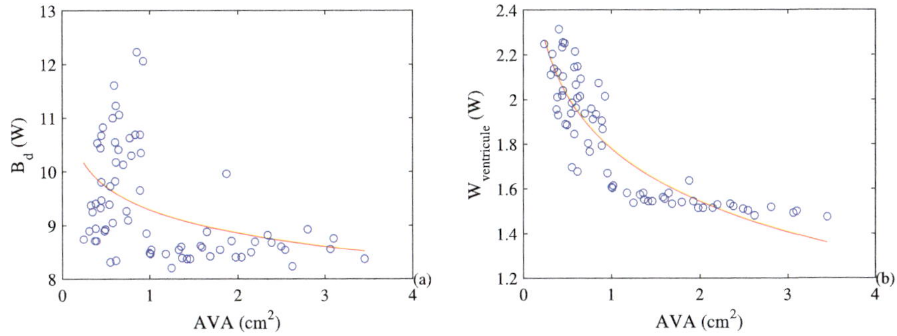

Fig. 6.7 **a** Exergy destroyed by the heart (W) and **b** Power realized in the heart (W) due to the area of the aortic valve (cm^2). *Source* [1]

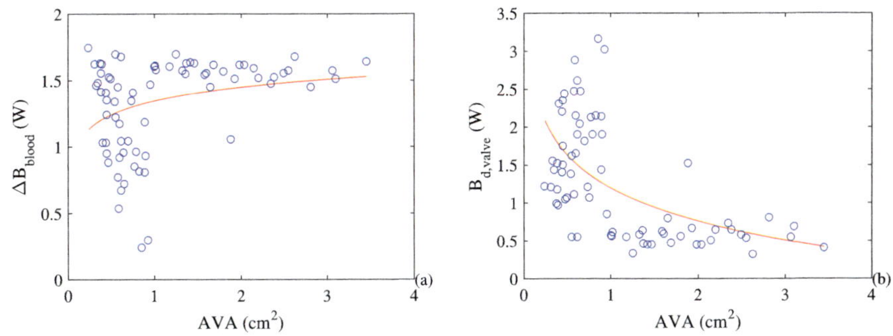

Fig. 6.8 **a** Exergy variation of the currents entering and leaving the heart (W) and **b** exergy destroyed only in the valves (W) due to the area of the aortic valve (cm^2). *Source* [1]

[1], but with the same order of magnitude as the values shown in the figure. From this point on, a more detailed study is needed regarding internal inefficiencies, including the destruction of exergy due to the blood flow through the valves. However, it should be noted that the executed power increases with the severity of the stenosis.

Now, Fig. 6.8 highlights that the greater the blood flow (lower severity of stenosis), the greater the variation in blood exergy, while the exergy destroyed only in the valve decreases (Fig. 6.8b). It is important to note that in Fig. 6.8a, the variation in blood exergy is a product of the entire heart process. In other words, this increase in exergy is what ensures circulation. It is expected that it will be larger when the irreversibilities associated with the increase in stenosis severity (pressure drop across the valves) are smaller. At the same time, Fig. 6.8b demonstrates the increase in exergy destruction in the heart solely related to the increase in pressure drop across the valves.

Regarding nutrient consumption, Fig. 6.9 demonstrates that nutrient consumption in metabolism increases as the area of the aortic valve decreases. This is partly due to the increase in irreversibilities intrinsic to human metabolism.

Figure 6.10 shows the results of the two definitions of exergetic efficiency given by Eqs. 6.14 and 6.15. Figure 6.10a describes the result of the power-to-metabolic exergy

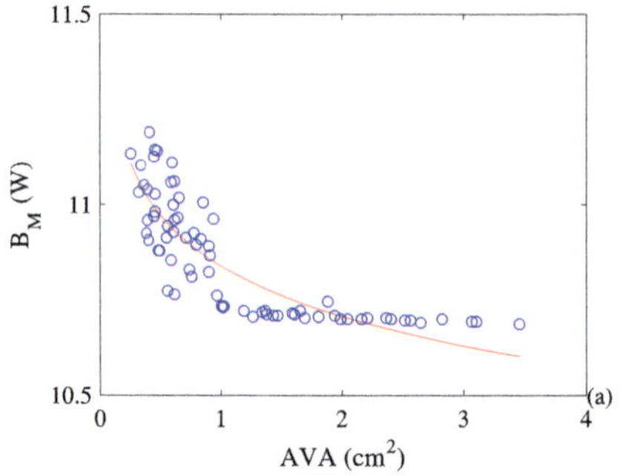

Fig. 6.9 **a** Exergy associated with metabolism (W) and **b** Exergy associated with heat released by metabolism (W) due to the area of the aortic valve (cm^2). *Source* [1]

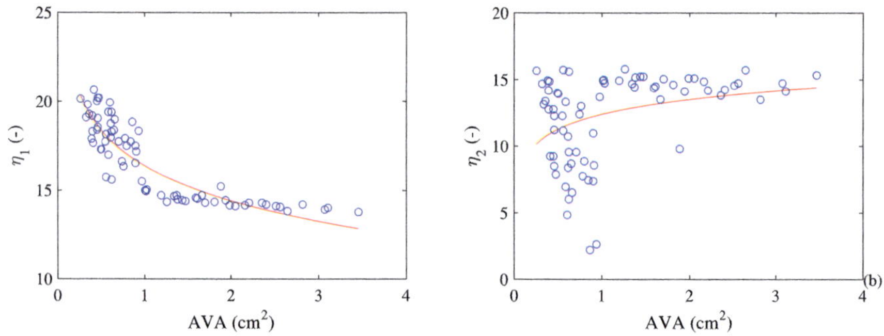

Fig. 6.10 **a** Exergy efficiency of the heart according to Eq. 6.14 and **b** exergy efficiency of the heart according to Eq. 6.15 as a function of the area of the aortic valve (cm^2). *Source* [1]

ratio. This efficiency corresponds to the traditional First Law efficiency commonly described in medical literature, as mentioned in [3]. The purpose is to identify work as the useful output. Now, the Second Law of Thermodynamics provides additional insights. By considering the heart as a control volume, the exergy input is the electrical signal that causes the heart to move, that is, the power executed by the heart would be the exergy input. Thus, Fig. 6.10b shows how much of the input exergy actually results in a variation of the flow exergy. The result is a different behavior, where the efficiency calculated by Eq. 6.14 increases with the severity of the stenosis, while the efficiency calculated by the definition of Eq. 6.15 decreases with the severity of the stenosis. Therefore, it can be concluded that the useful output of the heart would be the increase in the exergy of the outflow blood with respect to the inflow.

One final discussion concerns the nature of the work performed by the heart. Given the results illustrated in Fig. 6.10, it can be stated that work is not a useful output of the heart, and further analysis is needed to understand its nature. Whether work is considered an inflow or outflow of exergy in the heart is a matter that requires further investigation.

6.3 Concluding Remarks

In this chapter, a model of a heart was proposed taking into account various interactions between the cardiovascular system and the surrounding environment, i.e., the body. The results show similarity with those previously published in the literature of the field, such as [18]. One point to be emphasized in this analysis is the use of experimental results from the literature, as well as obtaining data from the medical literature on both healthy patients and those with severe stenosis. As a limitation, it should be noted that there is a lack of experimental results in the medical literature regarding healthy individuals. This is most likely due to the lack of interest in conducting in-depth studies on this particular patient group. Therefore, it was necessary to rely on data from cardiac prostheses. This result opens up possibilities for analyses involving the hydrodynamics of the heart, as studied by [19]. From this analysis, it is possible to conclude that:

- Both metabolism and cardiac work increase with the severity of stenosis. This result explains a possible hypertrophy that occurs in the heart affected by the pathology.
- At the same time, the variation of blood exergy decreases with the increase of AVA restriction (the quality of energy variation decreases with the severity of the pathology). This results in an increased cardiac effort, leading to higher losses and higher values of exergy destruction as the pathology progresses.
- These results can be used in the medical field in an attempt to understand the physiological responses of the heart caused by the pathology. In this scenario, the focus is on aortic valve stenosis. The main idea would be the use of exergy efficiency as an indicator of the degree of severity of the pathology, as its behavior is a consequence of the presence of the heart disease. Therefore, changes in blood hemodynamics and valves cause changes in the second law of thermodynamics that are not observed with the use of only the first law thermodynamics.

References

1. Roll JB, Borges ML, Mady CEK, Oliveira-Junior S (2019) Exergy analysis of the heart with a stenosis in the arterial valve. Entropy 21:563
2. Henriques IB, Mady CEK, Oliveira-Junior S (2016) Exergy model of the human heart. Energy 117:612–619
3. Guyton AC, Hall JE (2011) Guyton and Hall textbook of medical physiology. Elsevier
4. Steendijk P, Brinke EA (2008) Invasive assessment of cardiac efficiency. Heart Metabol 39:33–36
5. Schaldach M, Hutten H (1992). New concepts in electrotherapy of the heart. In: 1992 14th annual international conference of the IEEE engineering in medicine and biology society 2, pp 705–706
6. Blick EF, Stein PD (1977) Work of the heart: A general thermodynamics analysis. J Biomech 10(9):589–595

7. Suga H, Hayashi T, Shirahata M (1981) Ventricular systolic pressure-volume area as predictor of cardiac oxygen consumption. American J Physiol-Heart Circulatory Physiol 240(1):H39–H44
8. Izzi G, Zile MR, Gaasch WH (1991) Myocardial oxygen consumption and the left ventricular pressure-volume area in normal and hypertrophic canine hearts. Circulation 84(3):1384–1392
9. Takaoka H, Takeuchi M, Odake M, Yokoyama M (1992) Assessment of myocardial oxygen consumption (Vo2) and systolic pressure-volume area (PVA) in human hearts. Eur Heart J 13:85–90
10. Takaoka H, Takeuchi M, Odake M, Hayashi Y, Hata K, Mori M, Yokoyama M (1993) Comparison of hemodynamic determinants for myocardial oxygen consumption under different contractile states in human ventricle. Circulation 87(1):59–69
11. Mady CEK, Oliveira-Junior S (2013) Human body exergy metabolism. Int J Thermodyn 16(2):73–80
12. Gorlin R, Gorlin SG (1951) Hydraulic formula for calculation of the area of the stenotic mitral valve, other cardiac valves, and central circulatory shunts. I. American Heart J 41(1):1–29
13. Bache RJ, Wang Y, Greenfield-Junior JC (1973) Left ventricular ejection time in valvular aortic stenosis. Circulation 47(3):527–533
14. Klingensmith ME (Ed) (2008) The Washington manual of surgery. Lippincott Williams & Wilkins
15. Rosenhek R, Binder T, Maurer G, Baumgartner H (2003) Normal values for Doppler echocardiographic assessment of heart valve prostheses. J American Soc Echocardiogr 16(11):1116–1127
16. Zoghbi WA, Chambers JB, Dumesnil JG, Foster E, Gottdiener JS, Grayburn PA, Zabalgoitia M (2009) Recommendations for evaluation of prosthetic valves with echocardiography and doppler ultrasound. J American Soc Echocardiogr 22(9):975–1014
17. Sordelli C, Severino S, Ascione L, Coppolino P, Caso P (2014) Echocardiographic assessment of heart valve prostheses. J Cardiovascular Echogr 24(4):103–113
18. Çatak J, Özilgen M, Olcay AB, Yılmaz B (2018) Assessment of the work efficiency with exergy method in ageing muscles and healthy and enlarged hearts. Int J Exergy 25(1):1–33
19. Abdala L, Mady CEK, Correa MR (2025) A Heart Chamber Model Using a Capacitance Function Combined With the Navier-Stokes Equations. Math Methods Appl Scie 48(6):6447–6468

Chapter 7
Respiratory System

7.1 Introduction

Based on Albuquerque et al. [1], Henriques et al. [2] proposed a model of the respiratory system of the human body using the thermodynamic properties and metabolic reactions previously obtained by [3]. It is noteworthy that the entire respiratory system is within the boundaries of Fig. 3.1. The major changes in this work compared to the literature include the consideration of work performed by the lungs, the calculation of lung metabolism, and the inclusion of a more complex composition of air (O_2, CO_2, N_2, CO), among other factors. Therefore, with such a model, it becomes possible to perform exergy analysis in the respiratory system of the human body, as well as demonstrate the evolution of the models by Albuquerque et al. [1, 4] Albuquerque and Yanagihara [5], Henriques et al. [2], Cenzi et al. [6, 7], with the increased understanding of biological phenomena.

This control volume corresponds to a section of the phenomenological model proposed by [4, 5], where compartments represent different body parts. The blood enters the lung capillaries from the venous compartment, exchanging mass with gases from the pulmonary compartment. It is directed to the arterial compartment and subsequently to the tissue compartments, where it exchanges mass with the different tissues. Therefore, in Chap. 3 the central reservoir receives venous blood from all the tissues Fig. 5.2 and leaves as arterial blood. "The pump" (heart) was described in Chap. 6 in order for the blood to be delivered to the cells. Herein, we describe how the gas exchanges occur in the cells and between blood and lungs. Figure 7.1 represents this model as well as the representation of the compartments. It is important to note that there is a shunt in the capillaries to keep the blood pH within a specific value range.

It is worth noting that carbon monoxide is produced in human metabolism, but at a very small rate that does not cause any harm. Exogenous CO is toxic because it has an affinity for hemoglobin that is 200 to 300 times greater than that of O_2. Therefore, even relatively small concentrations of inhaled CO can reduce the oxy-

C. E. Keutenedjian Mady, *Exergy Analysis*, Green Energy and Technology,
https://doi.org/10.1007/978-3-032-01646-1_7

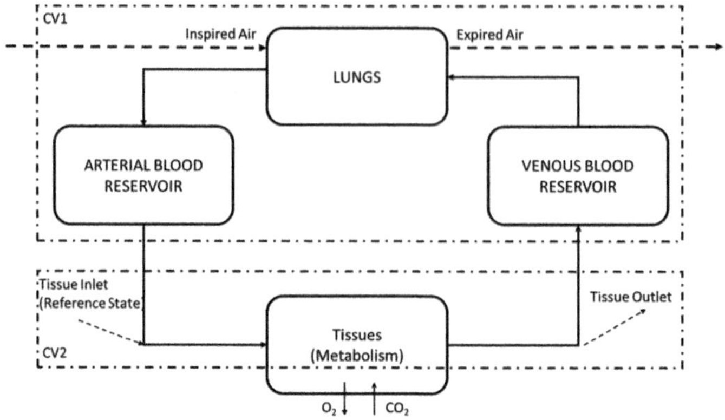

Fig. 7.1 Control volume used to perform the phenomenological model of the respiratory system. *Source* [6] adapted from [4]

gen released to the tissues, causing hypoxia. The toxicity of CO, therefore, is not caused by any direct harmful effect, but rather by occupying spaces that originally belonged to O_2 due to the increase in carboxyhemoglobin, which can lead to death. Since the reaction between carbon monoxide and hemoglobin is reversible, the treatment for CO poisoning involves reducing its inhaled concentration to zero, creating a maximum partial pressure gradient, increasing alveolar ventilation, and increasing the partial pressure of O_2, as pointed out in [8]. The concentrations of carboxyhemoglobin can be around values below 1%, where no observable effects are present in the body, while concentrations of 2%–5% can lead to alterations in the central nervous system (CNS), decreased ability to distinguish time and space, impaired visual acuity, changes in motor function, among others. Such factors are discussed by Cenzi et al. [6, 7].

The authors [2] and [6] proposed some modifications in the control volume of Fig. 7.1. In this new model (Fig. 7.2), the complete respiratory system is inside the control volume. The main differences here are the consideration of work done to the lungs ($\dot{W}_{res}$), the calculation of the lung metabolism ($\dot{M} \approx B_M$), and that the air has a more complex composition. Figure 7.2 is the control volume used in the present work to apply exergy analysis. The first exergy analysis model discussed in this paper has its foundations in this work. In this figure, the air ($\dot{B}_{a,in}$) enters the respiratory system with some exergy if the environment is considered the reference one proposed by [9], or with no exergy if the reference is the actual environment that the person is within. After the mass exchanges with the blood, the air leaves the respiratory system ($\dot{B}_{a,ex}$) with a higher temperature and humidity ratio. The blood enters as venous blood, rich in CO_2 (poor in O_2), and with exergy ($\dot{B}_{bl,v}$) and leaves as arterial blood, rich in O_2 ($\dot{B}_{bl,a}$).

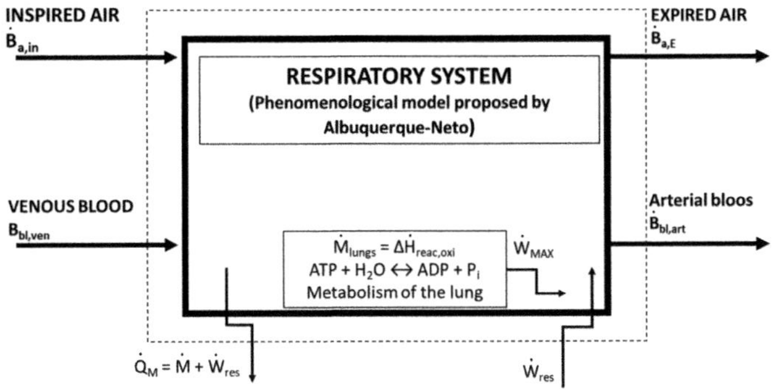

Fig. 7.2 Control volume to evaluate the exergy performance for the lungs under different conditions. *Source* [6] based on [4]

In this way, it is possible to calculate the exergy destruction rate in the respiratory system using Eq. 7.1. In literature, there are some articles that perform second law into the respiratory system such as Dutta and Chattopadhyay [10], Banerjee et al. [11].

$$\dot{B}_{dest}^{resp} = \dot{B}_{M_{lung}} + \dot{B}_{bl,v} + \dot{B}_{a,in} + \dot{W}_{resp} - \dot{B}_{bl,a} - \dot{B}_{a,ex} - \dot{B}_{Q_{M,lung}} \tag{7.1}$$

In Eq. 7.1, the term $\dot{B}_{dest}^{resp}$ refers to the exergy destruction during respiration in the lung, $\dot{B}_{M_{lung}}$ refers to the exergy metabolism of the lung, $\dot{B}_{bl,v}$ and $\dot{B}_{bl,a}$ refer to the exergies of venous and arterial blood, respectively; and $\dot{B}_{a,in}$ and $\dot{B}_{a,ex}$ refer to the exergies of inhaled and exhaled air, respectively. The term $\dot{B}_{\dot{Q}_{M,lung}} = \dot{Q}_{M,lung}(1 - T_0/T_{body})$ is the exergy associated with heat transfer from the lung to the rest of the body, obtained from Eq. 7.2. Finally, the term $\dot{W}_{resp}$ refers to the work performed by the intercostal muscles on the lung (against the lung) and therefore, it is added in Eq. 7.2.

Assuming steady-state conditions, the heat transferred by the lungs is the sum of its metabolism and the work done by the intercostal muscles and the diaphragm to the lungs, as in Eq. 7.2

$$\dot{B}_{Q_{M,pulm}} = \dot{Q}_{M,pulm}\left(1 - \frac{T_0}{T_{Lung}}\right) = \left(\dot{M}_{pulm} + \dot{W}_{resp}\right)\left(1 - \frac{T_0}{T_{Lung}}\right) \tag{7.2}$$

The work performed by the intercostal muscles, $\dot{W}_{resp}$, can be obtained according to Eq. 7.3, derived by [12]. In this equation, $\dot{\forall}_E$ represents pulmonary ventilation, and A_d is the skin area calculated using the classical equation proposed by Dubois [13].

$$\dot{W}_{resp} = 0.0755 \times A_d \times exp\left(0.196\frac{\dot{\forall}_E}{A_d}\right) \tag{7.3}$$

7.2 Simulations Procedure

7.2.1 *Model 1: Considering the Gases Dissolved in the Blood as Ideal*

The simplification of ideal gas allows for the evaluation of the exergy efficiency of the lung according to Eq. 7.4. The definition is based on the discussions proposed by [9], in which in chemical processes or processes involving separations and mixtures (where there is no well-defined output product) the exergy efficiency is defined as the ratio of the exergy leaving the control volume (in this case, the lung) to the exergy entering the control volume.

$$\eta_{ex} = \frac{\dot{B}_{a,ex} + \dot{B}_{bl,a} + \dot{B}_{Q_{M,lung}}}{\dot{W}_{resp} + \dot{B}_{M_{lung}} + \dot{B}_{bl,v} + \dot{B}_{a,in}} \tag{7.4}$$

In this equation, $\dot{B}_{bl,i}$ represents the exergy of blood, where the subscript i refers to arterial or venous blood. The exergy of the blood is calculated according to Eq. 7.5, where the exergy of the blood is divided by the exergy of the liquid and the exergy of the dissolved gases, as indicated in Eqs. 7.6 and 7.7.

$$\dot{B}_{bl} = \dot{B}_{liq,i} + \dot{B}_{O_2} + \dot{B}_{CO_2} + \dot{B}_{CO} \tag{7.5}$$

$$\dot{B}_{liq,i} = \dot{m}_{bl} c_{bl} \left(T_{bl,i} - T_0 - T_0 ln \frac{T_{bl,i}}{T_0} \right) \tag{7.6}$$

$$\dot{B}_g = \dot{m}_g \left[c_{p,g} \left(T_g - T_0 - T_0 ln \frac{T_g}{T_0} \right) + R_g T_0 ln \frac{P_g}{P_{g,0}} \right] \tag{7.7}$$

7.2.2 *Model 2: Considering the Gases Dissolved in the Blood and Bounds to Hemoglobin*

The change in exergy for any reaction can be evaluated as the Gibbs free energy change for the same reaction, but with the sign reversed, assuming it occurs at T_0. Therefore, in any reaction $\Delta b = -\Delta g$, with the requirement that one of the boundaries is at the environment temperature, the Gibbs function can be written as $g = h - T_0 s$. The results proposed by [14] and [15] are used to calculate the Gibbs free energy change of the oxygen transport process by hemoglobin. It is worth noting that these data are used because of the difficulty in evaluating the exergy of compounds in the human body as discussed in [3]. Equation 7.8 relates the equilibrium constant $k_{eq}^{'0}$ com $\Delta_r g^{'0}$, where the subscript r refers to a reaction. As pointed out in Sect. 3.3.2, the term $0'$ refers to the biological reference standard of the thermodynamic properties, being T_0 and P_0 temperature and pressure at environmental temperature (similar to biological one). The equilibrium constant, $k_{eq}^{'0}$, is also evaluated under these conditions.

$$\Delta_r g^{'0} = -\bar{R}T_0 \sum ln\left(k_{eq}^{'0}\right) \tag{7.8}$$

Equation 7.8 allows for the incorporation of the effect of O_2 and CO being transported by hemoglobin molecules in exergy analysis, rather than being solely dissolved in the blood as a perfect gas. This mode of reaction can be interpreted as an evolutionary aspect to maintain the concentration gradient between the gases present in the inhaled air and the blood passing through the alveoli. The binding form between O_2 and hemoglobin is called cooperative, meaning that the occurrence of the first binding facilitates the second, and so on successively. Therefore, in this text, it will be considered that hemoglobin is in the oxygenated or free form. Eq. 7.9 demonstrates how to evaluate the Gibbs change function in the reaction with O_2.

$$\Delta_r g^{'0} = \Delta_f g^{'0}\left[Hb\left(O_2\right)\right] - \Delta_f g^{'0}\left[\left(O_2\right)\right] - \Delta_f g^{'0}\left[Hb\right] \tag{7.9}$$

In this work, $\Delta_f g^{'0}[Hb]$, Gibbs variation of hemoglobin formation, is considered to be zero as discussed in [14] and [15] because it is the most stable allotropic form of this compound within the human body. The term $\Delta_f g^{'0}[O_2]$, on the other hand, corresponds to 16.1 kJ/mol as calculated in [18]. The remaining step is to calculate the Gibbs free energy changes for the four O_2 and hemoglobin bindings. Each one has its exergy release or absorption related to the binding and release of the hemoglobin.

To calculate the $\Delta_f g^{'0}[Hb(O_2)]$ term, it is necessary to evaluate the equilibrium constants of each binding, as shown in Table 7.1, obtained from [19]. The values represented in the table refer to both CO and O_2. It should be emphasized that, with those constants, it is possible to obtain the Gibbs free energy change as per Eq. 7.10.

$$\Delta_f g^{'0}\left[Hb\left(O_2\right)\right]_i = -RT_0 ln\left(k_{eq,i}^{'0}\right) + \Delta_f g^{'0}\left[Hb\left(O_2\right)\right]_{i-1} + \Delta_f g^{'0}\left[\left(O_2\right)\right] \tag{7.10}$$

By applying Eq. 7.10, it is possible to evaluate the values of Gibbs free energy change for each reaction, in order to apply exergy analysis as shown in Eq. 7.14 with the addition of this term.

Table 7.1 Equilibrium constant of successive reactions between hemoglobin and CO and O_2, obtained from [19]

Equilibrium constant	CO	O_2
$k_{eq,1}^{'0}$	4.62 x 10^6	1.96 x 10^6
$k_{eq,2}^{'0}$	1.61 x 10^6	1.25 x 10^6
$k_{eq,3}^{'0}$	2.50 x 10^8	2.68 x 10^7
$k_{eq,4}^{'0}$	4.64 x 10^8	1.52 x 10^9

Table 7.2 Gibbs free energy for fetal hemoglobin in (kJ/mol). Based on systematization proposed by [6] for biological reference conditions

	CO [kJ/mol]	O_2 [kJ/mol]
$\Delta_f G'^o(Hb(E))$	–150.09	–18.34
$\Delta_f G'^o(Hb(E))_2$	–301.04	–37.41
$\Delta_f G'^o(Hb(E))_3$	–464.96	–59.86
$\Delta_f G'^o(Hb(E))_4$	–634.09	–95.89

Model 2 was based on a modification proposed by [6] to include terms related to the exergy release (or absorption) associated with the binding of O_2 and CO with hemoglobin ($\Delta B_{Hb,i} = -\Delta g_{Hb,i}$). These values are summarized in Table 7.2. The majority of CO_2 is transported as bicarbonate ions in the blood, following a dissociation reaction. A detailed description of the calculations for those terms can be found in [6, 18, 19].

Hemoglobin binding with CO and O_2 is cooperative, meaning that once the first ligation happens, the following ones occur more easily. Following that, a hemoglobin molecule is commonly unbound or bound with four other molecules.

From Table 7.2, it is possible to calculate the variation of blood exergy. It has been considered that hemoglobin will either be unbound or completely bound to either O_2 or CO. It is important to recall that the first binding of the element E (O_2 or CO) to hemoglobin facilitates the subsequent three. Therefore, the variation of the exergy associated with the binding to hemoglobin is calculated according to Eq. 7.11. Here, $n_{art,E}$ refers to the amount of element E (O_2 or CO) bound to hemoglobin in arterial blood, and $nven, E$ refers to the amount of element E bound to hemoglobin in venous blood.

$$\Delta \dot{B}_E = -\left(\frac{n_{art,E} - n_{ven,E}}{4}\right) \Delta_f g'^0 \, [Hb\,(E)]_4 \tag{7.11}$$

In the case of CO_2, it is considered that the largest portion is transported in the form of bicarbonate ions, as shown in Eq. 7.12. It is important to point out that CO_2 is also transported by hemoglobin (approximately 10%), but this form has not been considered in this text.

$$H^+ + HCO_3^- \longleftrightarrow CO_2 + H_2O \tag{7.12}$$

Based on the thermodynamic properties from [18], it is possible to calculate the Gibbs free energy change for Reaction 7.12, resulting in –44.718 kJ/mol. Thus, it becomes possible to evaluate not only the difference in exergy of gases dissolved in the blood $\dot{B}_{a,ex} - \dot{B}_{BL,v}$ but also the exergy associated with the reactions, as shown in Eq. 7.13. Finally, the exergy analysis is expressed by Eq. 7.14.

$$\Delta \dot{B}_{reac} = \Delta \dot{B}_{O_2,reac} + \Delta \dot{B}_{CO_2,reac} + \Delta \dot{B}_{CO,reac} \tag{7.13}$$

From Eq. 7.13, a problem arises regarding the calculation of exergy efficiency. It is no longer possible to calculate the exergy flows entering and leaving the control volume separately, resulting in the results shown in Fig. 7.9, representing only the destroyed exergy.

$$\dot{B}_{dest}^{lung} = \dot{B}_{M_{lung}} + \dot{B}_{bl,v} + \dot{B}_{a,in} + \dot{W}_{lung} - \dot{B}_{bl,a} - \dot{B}_{a,ex} - \dot{B}_{Q_{M,lung}} - \Delta\dot{B}_{reac} \tag{7.14}$$

7.2.3 Model 3: Considering the Irreversibilities of Mass Exchange

In this section, only the mass transfer associated with gas diffusion between the two compartments will be considered. The exchanges occur between the air in the pulmonary alveoli and the blood, and their irreversibilities are associated with the difference in chemical potential between the two spaces. For CO_2, the variation in entropy is given by Eq. 7.15.

$$\Delta\dot{S}_{CO_2} = -\left(\dot{m}_{CO_{2,v}} - \dot{m}_{CO_{2,a}}\right) R_{CO_2} ln\left(\frac{P_{CO_{2,ex}}}{P_{CO_{2,a}}}\right) \tag{7.15}$$

The concentration of this gas is higher in venous blood, and its direction is from the blood to the alveolus. For O_2 and CO, the concentration is higher in the blood, and the gases move from the blood to the alveoli. Equations 7.16 and 7.17 represent the entropy change associated with this transport.

$$\Delta\dot{S}_{O_2} = -\left(\dot{m}_{O_{2,v}} - \dot{m}_{O_{2,a}}\right) R_{O_2} ln\left(\frac{P_{O_{2,ar}}}{P_{O_{2,exp}}}\right) \tag{7.16}$$

$$\Delta\dot{S}_{CO} = -\left(\dot{m}_{CO_v} - \dot{m}_{CO_a}\right) R_{CO} ln\left(\frac{P_{CO_{exp}}}{P_{CO_{ar}}}\right) \tag{7.17}$$

Therefore, by applying the Second Law of Thermodynamics at the boundary where the mass transfer occurs, it is possible to calculate the entropy generation associated with this mechanism, as well as the rate of exergy destruction using the Gouy-Stodola Theorem, as stated in [20]. Thus, $\dot{B}_d = T_0 \times \dot{\sigma}$

$$\dot{\sigma} = \Delta\dot{S}_{CO} + \Delta\dot{S}_{O_2} + \Delta\dot{S}_{CO_2} \tag{7.18}$$

7.3 Different Environments and Model Conditions for Simulations

7.3.1 Altitude and Exergy Behavior of the Lungs

The model used in the simulations procedure is proposed by Albuqerque and Yanagihara [5], where the mechanism of heat and mass exchanges is validated for the human lungs and the thermal system of the whole body. It is important to highlight that no modifications were necessary until this point.

The exergy balance was applied to different scenarios, resulting in exergy destruction rates as a function of altitude, acclimatization time (hemoglobin level in the blood), and carbon monoxide concentration in the atmospheric air. Four altitude levels were considered: 0, 1500, 3000, and 4500 m, which are associated with different barometric pressure levels, as shown in Table 7.3.

The concentration of hemoglobin in the blood as a function of time and altitude is given in Table 7.4, obtained from [2]. The amount of hemoglobin increases with altitude compared to that at sea level. This is a physiological response to the decrease in oxygen in the atmospheric air. The ventilation rate is another form of acclimatization, which has been discussed in studies such as [16] and [17]. This mechanism increases significantly in the first few minutes and continues to increase up to 28 days of exposure. From that point onwards, it becomes constant. Those values can decrease by half if the exposure period is on the order of years, being more prominent at altitudes higher than 1500 m. Therefore, it was assumed that ventilation is constant and equal when a person is exposed to 0 and 1500 m. Data is available for an altitude of 4500 m, as pointed out in [17], so a linear interpolation was performed for 3000 m.

7.3.2 Exergy Behavior of a Pregnant Woman and Fetus Under Intoxication of Carbon Monoxide

Maternal arterial blood exits the respiratory system (lungs) and is pumped by the heart [1] during left systolic ejection. From there, it travels to the placenta, where gas and nutrient exchanges occur, and leaves as venous blood. Subsequently, it passes through the heart before departing the systemic circulation and entering the

Table 7.3 Barometric pressure as a function of altitude

Altitude [m]	Pressure [kPa]
0	101.3
1500	84.5
3000	70.1
4500	57.7

Table 7.4 Concentration of hemoglobin in the blood as function of time and altitude. *Source* [6]

Alt. [m]	t [days]	[Hb] [g/dL]	$\dot{V}_{alveolar}$ [mL/min]
0	0	14.5 (reference value)	4212.0
0	2	14.5	4212.0
0	20	14.5	4212.0
0	60	14.5	4212.0
0	90	14.5	4212.0
0	0	17.03	4212.0
1500	0	14.5	5300.0
1500	2	14.17	5300.0
1500	20	15.66	5300.0
1500	60	15.66	5300.0
1500	90	15.66	5300.0
1500	0	17.03	5300.0
3000	0	14.5	5850.0
3000	2	13.82	6900.0
3000	20	16.28	7162.5
3000	60	16.33	7162.5
3000	90	16.53	7162.5
3000	0	17.03	7162.5
4500	0	14.5	6525.0
4500	2	13.32	7575.0
4500	20	16.78	9675.0
4500	60	16.83	9675.0
4500	90	17.03	9675.0

pulmonary circulation. Ultimately, the blood returns to the lungs to be oxygenated, though the work performed by the heart is not evaluated in the present analysis (as it was analyzed [1] in Chap. 6). Nevertheless, it is included in Fig. 7.3, which illustrates the gas exchange model for pregnancy. The figure represents the maternal respiratory system adapted from [4] and the gas exchanges in the placenta suggested in the present analysis.

The placenta is an organ that is responsible for performing the exchange of substances between the mother and the fetus during pregnancy. It plays the role of the lungs, intestines, and kidneys of the fetus. The placenta has a high metabolic rate, as evidenced by its high consumption of O_2 and glucose. It consists of two independent circulatory systems: the maternal circulation and the fetal circulation. Oxygenated blood from the uterine arteries enters the placenta, constituting a pool of oxygenated blood. Exchanges occur within this pool. Carbon dioxide and waste products are transported from the fetus through the umbilical arteries and passively diffuse into the maternal blood. Oxygen and nutrients, in turn, diffuse from the maternal blood

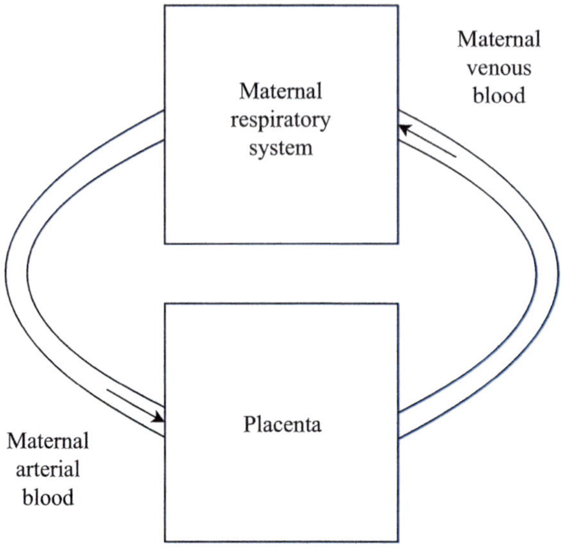

Fig. 7.3 Gas exchange model for pregnancy indicating the mother respiratory system obtained in [4] and the gas exchanges in the placenta proposed in the present analysis. *Source* [7]

Table 7.5 Parameters obtained in the literature representing the pregnant woman to change some physiological settings of the model of [4]

Parameter	Value	Reference
Cardiac output	7.2 $L.min^{-1}$	[21]
Blood volume	6.53 L	[22]
Oxygen consumption	249 $mL.min^{-1}$	[22]
Ventilation	8.51 $L.min^{-1}$	[22]
Hemoglobin concentration	12.1 $g.dL^{-1}$	[22]
Respiratory quotient	0.83	[22]

to the fetus [23]. The gas exchange process in the placenta can be modeled as a concurrent process. Table 7.6 presents key parameters related to umbilical blood flow.

The input data used in the model are summarized in Table 7.5, which represents physiological conditions during pregnancy with greater accuracy. These parameters were obtained from the study by [4].

The placenta is an organ that is responsible for performing the exchange between mother and fetus. It plays the role of the lungs, intestines, and kidneys of the fetus during the pregnancy. It has a high metabolism level, as is indicated by its high O_2 and glucose consumption. This organ has two independent circulatory systems: the maternal and the fetus. Arterial blood from the uterine arteries enters the placenta, constituting a lake of arterialized blood. It is in this lake that exchanges take

Table 7.6 Parameters of umbilical and fetal blood flow obtained in literature

Parameter	Value	Reference
Umbilical blood flow	66 mL/min.kg	[24]
Fetal O_2	6.53 L	[24]
Hemoglobin	14.5 g/dL	[24]
P_{O_2} venous umbilical blood	28.8 mmHg	[24]

place. Carbon dioxide and waste products come from the umbilical arteries and are exchanged by passive diffusion to the maternal blood. Oxygen and nutrients diffuse from the maternal blood to the fetus [23]. The gas exchange process in the placenta can be modeled as a concurrent process. Table 7.6 lists important parameters of umbilical blood flow.

Similarly to the approach followed by [25], the placenta was modeled as two compartments: one representing the maternal blood flow and the other representing the fetal blood flow. The pressure of gases within each compartment is assumed to be constant and equal to the pressure at which it exits the correspondent compartment. Maternal arterial blood enters the placenta with an oxygen pressure of (P_{mar}), 90 mmHg and exits as venous blood with a pressure of 40 mmHg, P_{mve}. Fetal arterial blood enters with an oxygen pressure (P_{far}) of 25 mmHg and exits as venous blood (P_{fve}) at 40 mmHg. The umbilical veins are named as such because, even though arterial blood is low in oxygen, it exits the fetus's heart. The rate of volumetric gas diffusion ($\dot{\forall}$) for a specific gas can be calculated using Eq. 7.19:

$$\dot{\forall} = D_L(P_{m,ve} - P_{f,ve}) \tag{7.19}$$

In this equation, D_L (mL/min.kg) is the diffusion coefficient and P_i (mmHg) is the partial pressure of the gas in the compartment. The mass flow rate (kg/s) at which the blood gains a gas is given by Eq. 7.20.

$$\dot{m}_{diffusion} = \dot{\forall}(C_{g,ar} - C_{g,ve}) \tag{7.20}$$

where $\dot{\forall}$ is the blood flow rate ($m^3.s^{-1}$) and C is the gas total concentration (kg/m^3) in the blood, which can be evaluated using Eq. 7.21 ([4, 25]).

$$C = \alpha_b P + N[Hb]S(P) \tag{7.21}$$

where S(P) is a function representing the oxygen dissociation curve, α is the solubility coefficient of the gas in the blood ($mL_{O_2}/(mL_{blood}.mmHg)$), [Hb] is the hemoglobin concentration (g/mL), and N is the total carrying capacity in the blood ($mL_{O_2}mL/mL_{bl}$).

To obtain the concentration of carbon monoxide in fetal venous blood under steady-state conditions, Eq. 7.22 was numerically solved using the bisection method until the pressure change was within a specified error, as defined in Eq. 7.23. It is important to note that 7.22 is a combination of Eqs. 7.19 and 7.20, and that the carbon monoxide pressure in the arterial fetal blood compartment is the same as the venous pressure in the last iteration, as indicated in (Eq. 7.24). These equations are described in [7].

$$D_L(P_{m,ve} - P^n_{f,ve}) = \dot{V}_b(C_{f,ve}(P^n_{f,ve}) - C_{f,a}(P^{(n-1)}_{f,ve}) \tag{7.22}$$

$$|(P^n_{f,ve} - P^{(n-1)}_{f,ve})/(P^n_{f,ve})| < \epsilon \tag{7.23}$$

$$P^n_{f,ar} = P^{(n-1)}_{f,ve}. \tag{7.24}$$

7.3.2.1 Model Modification

The analysis used for the maternal respiratory system is adapted for the placenta. Figure 7.4, shows how the placenta is modeled. Therefore, as a mass exchanger with its own metabolism (which is expected from an organ), it is important to bear in mind that the fetal tissue forms the placenta, and therefore, is an organ from the son.

The different analyses are described by the equations as commented in the next few paragraphs. While the phenomenological model takes into account the reaction between O_2, CO, and hemoglobin in [5], the early studies involving exergy analysis considered the mixture of blood and dissolved gases (O_2, CO, and CO_2) as ideal, without taking into account the effect of the reaction with hemoglobin. Such simplification is due to the difficulty in calculating the chemical exergy of a molecule like hemoglobin, which contains both organic and inorganic atoms in its chain. The

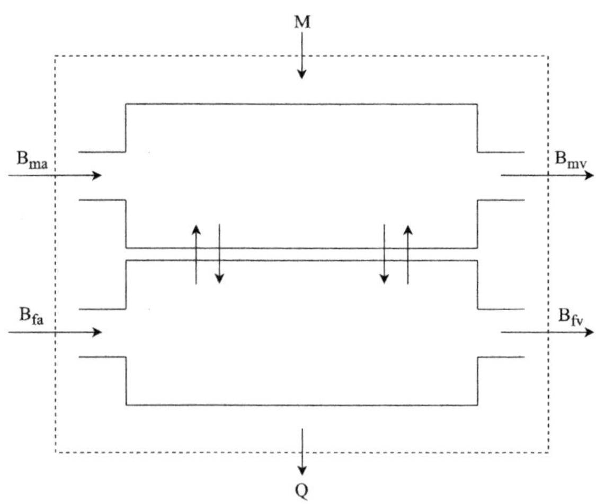

Fig. 7.4 Placenta with the stream of the mother and fetus arterial and venous blood. *Source* [7]

same difficulty is encountered in ATP, ADP, and P_i, as discussed by [3]. This section aims to show these first models, as in [1] and its evolution conducted by [6, 7].

- **Model 1**: In the following equations, m refers to maternal blood, while f to fetal. Also, the subscript ve refers to venous blood and ar to arterial. $\dot{B}_M$ is the exergy rate associated with the metabolism and $\dot{Q}$ is heat transferred from the placenta to it surroundings. T_0 is the reference temperature and T_c is the placenta temperature. The efficiency, based on the discussions carried out in [9] is defined as:

$$\eta_b = \frac{\dot{B}_{ve,m} + \dot{B}_{ve,f} + \dot{Q}_m(1 - T_0/T_c)}{\dot{B}_M + \dot{B}_{ar,m} + \dot{B}_{ar,f}} \tag{7.25}$$

Destroyed exergy rate is given by Eq. 7.26:

$$\dot{B}_d = \dot{B}_{ar,m} + \dot{B}_{ar,m} + \dot{B}_M - \dot{B}_{ve,m} - \dot{B}_{ve,f} - \dot{Q}(1 - T_M/T_c) \tag{7.26}$$

- **Model 2**: Based on a modification proposed by [6] to include terms related to the exergy release (or absorption) associated with the binding of O_2 and CO with hemoglobin ($\Delta B_{Hb,i} = -\Delta g_{Hb,i}$). These values are summarized in Table 7.2. The majority of CO_2 is transported as bicarbonate ions in the blood, following a dissociation reaction. A detailed description of the calculations for those terms can be found in [6, 18, 19]. Hemoglobin binding with CO and O_2 is cooperative, meaning that once the first ligation happens, the following ones occur more easily. Following that, a hemoglobin molecule is commonly unbound or bound with four other molecules. The exergy destruction rate will be described as follows by Eqs. 7.27–7.28

$$\dot{B}_d = \dot{B}_{ar,m} + \dot{B}_{ar,f} + \dot{B}_M - \dot{B}_{ve,m} - \dot{B}_{ve,f} - \dot{Q}(1 - T_0/T_c) + \Delta\dot{B}_{reac} \tag{7.27}$$

$\Delta\dot{B}_{reac}$ is given by

$$\Delta\dot{B}_{reac} = \Delta\dot{B}_{O_{2maternal}} + \Delta\dot{B}_{O_{2fetal}} + \Delta\dot{B}_{CO_{maternal}} + \Delta\dot{B}_{CO_{fetal}} \tag{7.28}$$

- **Model 3**: Focuses solely on the entropy generation associated with mass transfer by diffusion. This model does not consider other metabolic effects, heat transfer, or mass transfer with the rest of the body. These are in Eqs. 7.29–7.32

$$\Delta\dot{S}_{CO_2} = (\dot{m}_{CO_{2ve}} - \dot{m}_{CO_{2ar}})R_{CO_2}\ln\frac{P_{CO_{2m,ve}}}{P_{CO_{2f,ve}}} \tag{7.29}$$

$$\Delta\dot{S}_{CO} = -(\dot{m}_{CO_{ar}} - \dot{m}_{CO_{ve}})R_{CO}\ln\frac{P_{CO_{f,ve}}}{P_{CO_{m,ve}}} \tag{7.30}$$

$$\Delta\dot{S}_{O_2} = -(\dot{m}_{O_{2ar}} - \dot{m}_{O_{2ve}})R_{O_2}\ln\frac{P_{O_{2f,ve}}}{P_{O_{2m,ve}}} \tag{7.31}$$

In these equations, the pressure in each gas in the maternal (m) and fetal (f) blood is demonstrated, although just the values from venous (ve) blood are used. Then, the entropy generation is given by Eq. 7.32 and exergy destruction by $\dot{B}_d = T_0 \times \dot{\sigma}$.

$$\dot{\sigma} = \Delta \dot{S}_{CO_2} + \Delta \dot{S}_{CO} + \Delta \dot{S}_{O_2}. \tag{7.32}$$

7.4 Results and Discussions

Three different approaches to exergy analysis were applied to each system, following the methods described by [6]. The first method considers that the gases in the blood are carried dissolved with their partial pressures [2]. The second method takes into account that oxygen and carbon monoxide are also carried bound to the hemoglobin [6]. Finally, the second law was applied to evaluate the intrinsic entropy generation due to diffusion between regions with finite pressure gradients. The main objective of the study is to assess how carbon monoxide poisoning can affect the exergy behavior of the maternal lungs, which should be similar to the findings of [6]. Additionally, there is an interest in understanding the fetus's behavior for pregnant women, considering that the mother's blood undergoes significant alterations in environments with abnormal concentrations of CO. Such scenarios can occur in several locations within the urban ecosystem.

7.4.1 Effect of Altitude in the Carbon Monoxide Behavior of the Lungs

Figures 7.5, 7.6, 7.7 and 7.8 show the results of exergy efficiency and exergy destruction rate as a function of CO concentration and hemoglobin concentration, representing the time of acclimatization of individuals at higher altitudes. The exergy destruction rate decreases when the model is subjected to higher altitudes and undergoes acclimatization, and it returns to lower regions. At the same time, the exergy destruction rate decreases with increasing CO concentration in the environment (resulting in increased exergy efficiency). This result is unexpected from a medical perspective, since an increase in exergy destruction would be expected to indicate an increase in the harm associated with the gas. Considering only the mass transfer mechanism, an increase in exergy destruction would indicate that there is an exchange of mass with a higher concentration gradient, which indeed occurs. In Fig. 7.5 show only two curves of (a) the exergy efficiency and (b) the destroyed exergy rate, because there is only one return from 1500 m of altitude.

The same reasoning described for Fig. 7.5 can be applied to analyze Figs. 7.6, 7.7 and 7.8, but with a broader range of values since it takes into account that the model

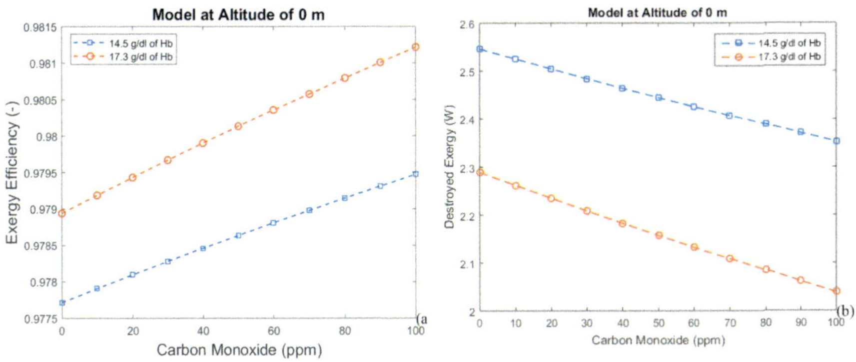

Fig. 7.5 **a** Exergy efficiency and **b** destroyed exergy as functions of carbon monoxide and hemoglobin (Hb) concentration at sea level. *Source* [6]

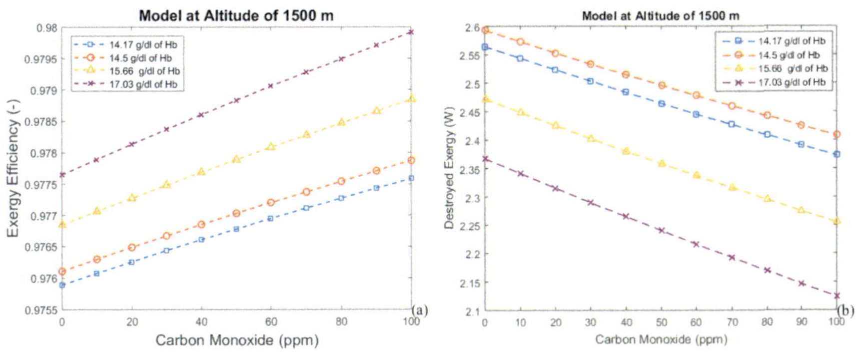

Fig. 7.6 **a** Exergy efficiency and **b** destroyed exergy as functions of carbon monoxide and hemoglobin concentration at 1500 m. *Source* [6]

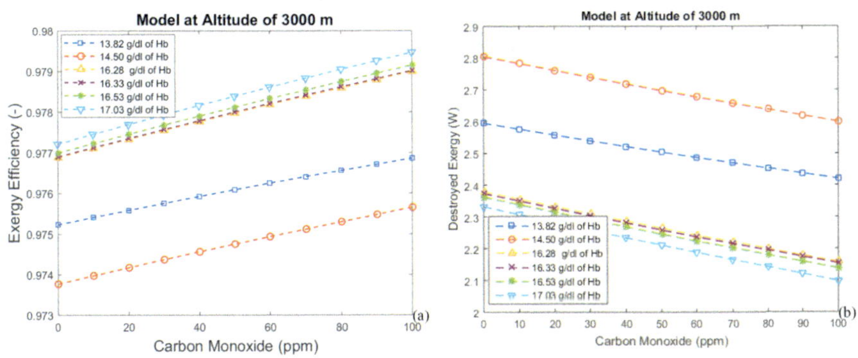

Fig. 7.7 **a** Exergy efficiency and **b** destroyed exergy as functions of carbon monoxide and hemoglobin concentration at 3000 m. *Source* [6]

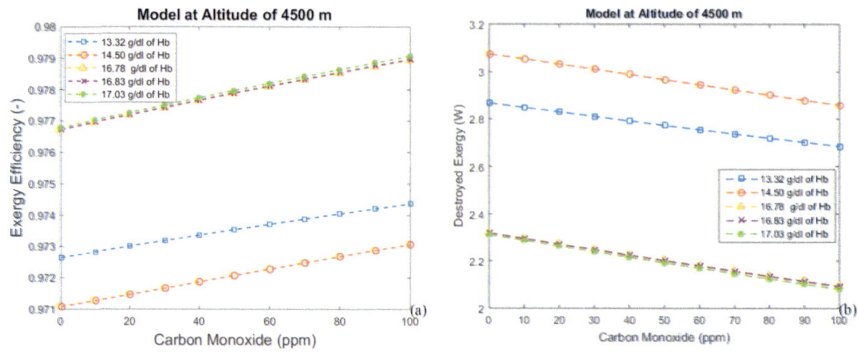

Fig. 7.8 **a** Exergy efficiency and **b** destroyed exergy as functions of carbon monoxide and hemoglobin concentration at 4500 m. *Source* [6]

underwent a certain incremental period of time for the increase in hemoglobin concentration and changes in alveolar ventilation, i.e., an acclimatization process. These two modifications are supposed to increase the mass transfer and diffusion inside the lungs. The increase in the alveolar ventilation causes an increase in metabolism and gas exchanges. The acclimatization increases the amount of hemoglobin in the blood, which intensifies the mass transfer in the lungs. The higher the hemoglobin concentration, the higher is the increase in the exergy efficiency and decrease of the destroyed exergy for a fixed amount of CO; nevertheless, one of the reasons that the destroyed exergy decreases with the increment of carbon dioxide intoxication is related to the fact that the body exchanges mass "better" when the gas is CO.

Figure 7.9a–d indicate the rate of exergy destruction as a function of the effect of carbon monoxide intoxication in the lungs and altitude for different hemoglobin concentrations. Figure 7.9a shows that at sea level (altitude of 0 m) and a hemoglobin concentration of 17.03 g/dL (individuals acclimatized to 4000 m), there is a difference of approximately 0.06 W compared to individuals acclimatized to sea level. The same reasoning can be applied to Fig. 7.9b, but this difference increases significantly for Fig. 7.9c, d, indicating that the higher the hemoglobin concentration, the greater the damage associated with carbon monoxide. It should be noted that the presence of hemoglobin in the blood is an evolutionary mechanism that maintains the pressure gradient in the alveoli. In other words, it is natural that the better the blood transports gases, the higher the exergy destruction will be, which is observed in these figures. That is, the greater the amount of hemoglobin, the more efficient the gas transport is, but with higher irreversibilities.

Furthermore, regarding the acclimatization period, it is worth noting that individuals who are less acclimatized to higher altitudes tend to have a lower rate of exergy destruction. There is also a change in behavior for the two lower values of hemoglobin concentration. Such a fact is associated with the increase in pulmonary ventilation in the initial moments when transitioning from a lower altitude to higher regions.

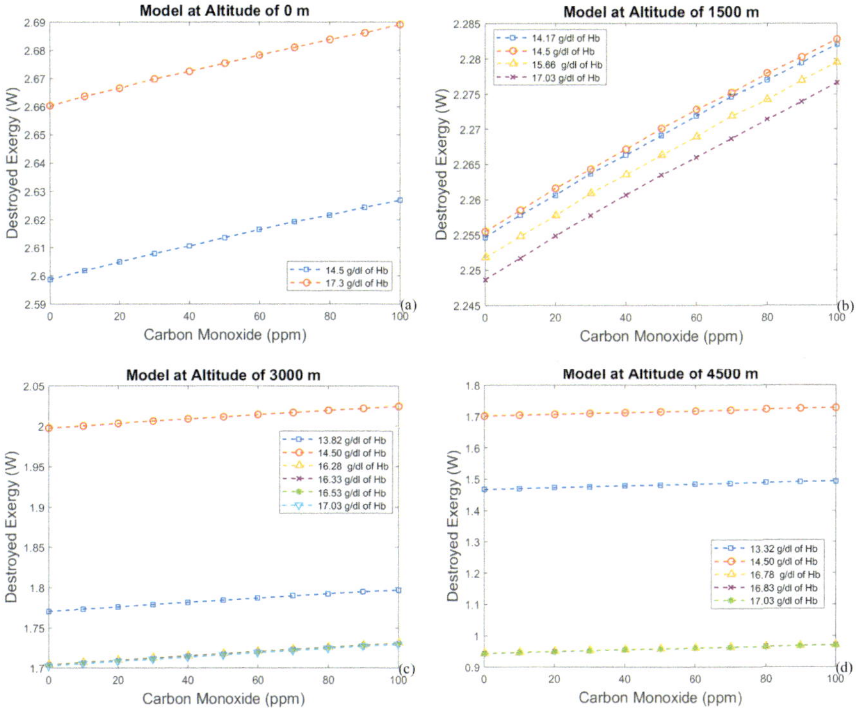

Fig. 7.9 Destroyed exergy as a function of carbon monoxide and hemoglobin concentration at **a** 0 m, **b** 1500 m, **c** 3000 m, and **d** 4500 m. *Source* [6]

By comparing the results from previous figures, considering the gases dissolved as ideal in the blood or reacted, it highlights the importance and necessity of comparing the mass transfer mechanism at different altitudes and their irreversibilities, as shown in Fig. 7.10. It is important to keep in mind that the phenomenological model proposed by Albuqerque and Yanagihara [5] takes into account the different forms of gas transport. Figure 7.10 shows that an increase in hemoglobin concentration is directly related to a decrease in exergy destruction (for the same concentration of CO in the air). Another result is the increase in exergy destruction related to the mass transfer rate, i.e., as atmospheric CO concentrations increase, the mass transport with a finite gradient and its irreversibilities increase. It is important to note that Fig. 7.10d presents a different trend from the others, which can be justified in two ways. The first possibility is that the model is outside the valid range, while the second possibility is related to the fact that all physiological changes are more pronounced at 4000 m, which alters the exergetic behavior of the lungs for different hemoglobin concentrations.

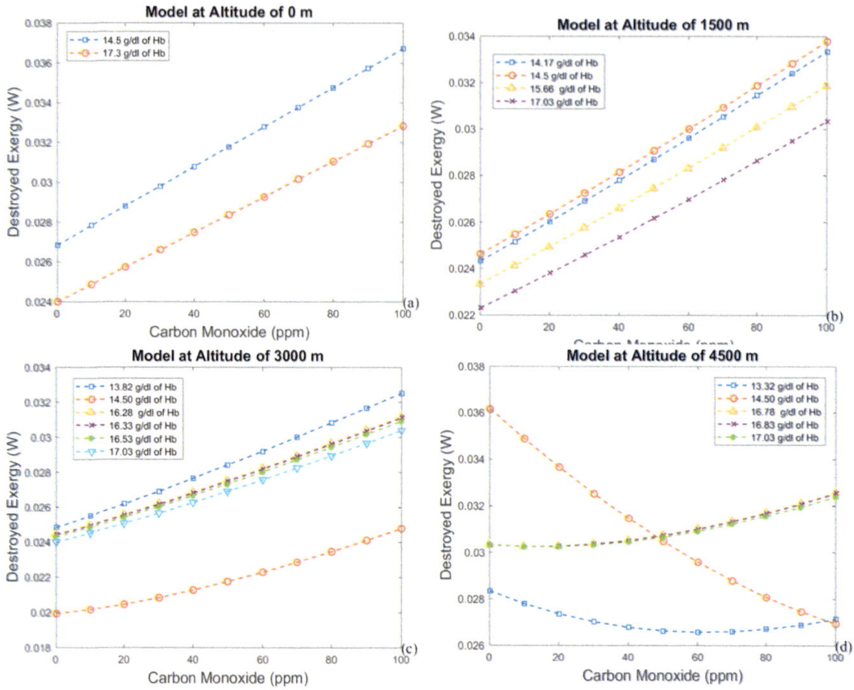

Fig. 7.10 Destroyed exergy as function of carbon monoxide and hemoglobin concentration at **a** 0 m, **b** 1500 m, **c** 3000 m, and **d** 4500 m. *Source* [6]

7.5 Results of the Exergy Behavior of Gas Exchanges in the Placenta During Pregnancy

Figure 7.11 depicts the partial pressure of carbon dioxide (CO_2) in maternal blood as a function of carbon monoxide concentration in the environment (CO–ppm). It can be observed that CO_2 level does not significantly change with increasing CO concentration. This is mainly due to the different mechanism of transport of CO_2. Carbon dioxide pressure is lower than that found in the literature, possibly due to the increase in ventilation that occurs in pregnancy, as discussed by [21].

Maternal exergy efficiency is presented in Fig. 7.12. It should be noted that this efficiency can only be evaluated for model 1, where the exergy of each stream can be defined, and therefore not only the change of exergy in hemoglobin binding as carried out in model 2. The efficiency increases with CO poisoning as found for a male in previous work by [6]. However, with significantly higher efficiency, that can be attributed to the enhanced ventilation that occurs during pregnancy.

Figure 7.13 shows the destroyed exergy rate for the maternal respiratory system evaluated using the three different models. In the first method, the results indicate a decrease in the exergy destroyed rate with carbon monoxide poisoning (which may

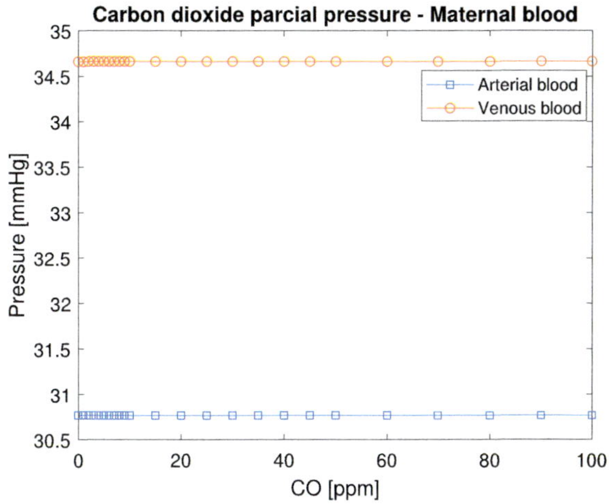

Fig. 7.11 CO_2 partial pressure in maternal blood. The conversion of CO in ppm to kg/m^3 is to divide by 0.001. *Source* [7]

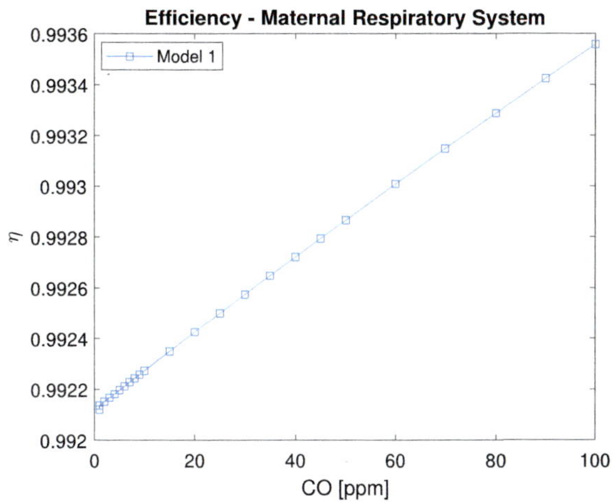

Fig. 7.12 Maternal respiratory system exergy efficiency. The conversion of CO in ppm to kg/m^3 is to divide by 0.001. *Source* [7]

be misleading when compared to the exergy efficiency, which increases with CO intoxication), while for the other techniques this trend grows; higher mass transfer increases the irreversibilities. In the third method, the exergy destruction rate is the smallest among the three methods, since it only considers diffusion and its irreversibilities, excluding metabolism and work. However, this method can still be used to support the development of the second method, as they agree in their general behavior. Moreover, the fact that most of the irreversibilities in the lungs are not related to the transport phenomena, but rather to other terms, such as metabolic rate, may indicate an evolutionary mechanism aimed at reducing this physical quantity.

Figure 7.14 illustrates the destroyed exergy rate per oxyhemoglobin percentage in arterial blood, allowing for a comparison with Fig. 7.13. It is possible to infer how blood saturation with O_2 changes as CO poisoning increases. Moreover, the saturation of O_2, often referred to as SO_2, is a commonly used index in medical

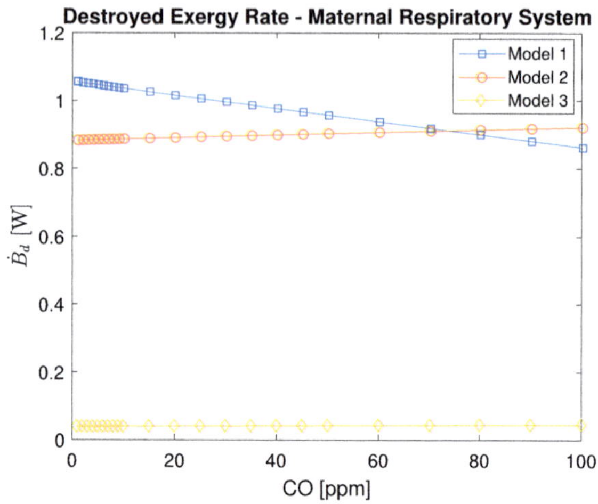

Fig. 7.13 Maternal respiratory system destroyed exergy rate as a function of carbon monoxide intoxication in ppm. The conversion of CO in ppm to kg/m^3 is to divide by 0.001. *Source* [7]

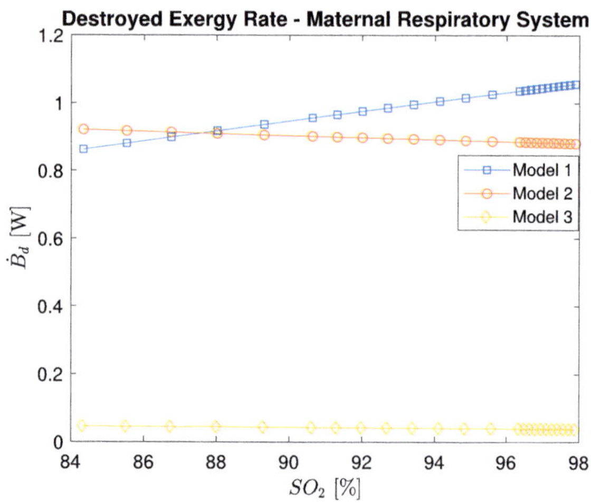

Fig. 7.14 Maternal respiratory system destroyed exergy rate per oxyhemoglobin (saturation of O_2 in the blood), which is a function of carbon monoxide poisoning. *Source* [7]

practice to assess the gas exchanges in the lungs (usually called lung capacity) when the person is under a respiratory infection [4, 26]. Lower values of SO_2 indicate the need for clinical intervention as they indicate the presence of hypoxia. Usually, in healthy individuals, the saturation is expected to be higher than 96% [26]. It is interesting to note that the curves of the destroyed exergy exhibit a negative slope, meaning that as the oxygen saturation increases from lower values (such as 84%) to higher values (96–98%), there is a healthier functioning of the lungs and lower concentrations of CO in the blood and environment.

On one hand, the model assumes that the pressure difference between maternal venous blood and fetal venous blood remains constant (Fig. 7.15) to ensure the diffusion of oxygen through the placenta from the mother to the fetus. It is evident from this figure that if the mother is intoxicated with levels as high as 100 ppm, the

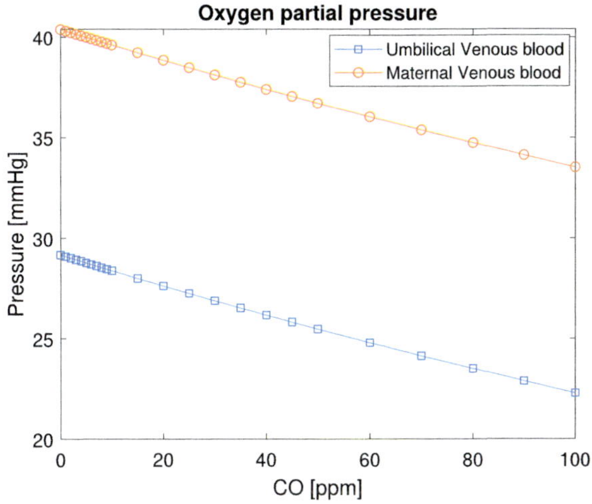

Fig. 7.15 O_2 partial pressure in maternal and fetal blood as a function of carbon monoxide intoxication in ppm. The conversion of CO in ppm to kg/m^3 is to divide by 0.001. *Source* [7]

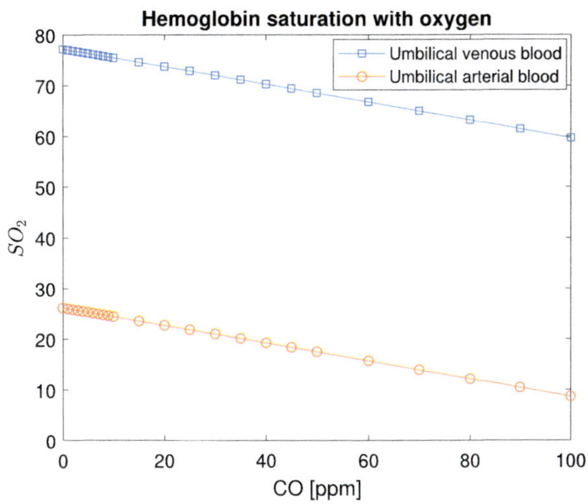

Fig. 7.16 Fetal hemoglobin saturation with oxygen as a function of carbon monoxide intoxication in ppm. The conversion of CO in ppm to kg/m^3 is to divide by 0.001. *Source* [7]

partial pressure of O_2 in the fetal blood will be significantly reduced, ranging between 20 and 25 mmHg. This condition is illustrated in Fig. 7.16. The oxygen saturation in the fetus varies from around 30% to 10%. This amount of oxygen may not be sufficient for the fetus's metabolism, potentially leading to hypoxia. On the other hand, Fig. 7.16 indicates that hemoglobin saturation with O_2 in fetal arterial blood would need to decrease to significantly low values in order to maintain a constant supply of oxygen (through fetal O_2 gradient with the mother's blood). Recalling that the hemoglobin saturation with carbon monoxide in fetal blood never reaches maternal levels (Fig. 7.17). Although it is assumed that the initial concentration of carbon monoxide is 2%, [27], even when the mother is not exposed to CO, it is the usual concentration found in the blood of a healthy fetus.

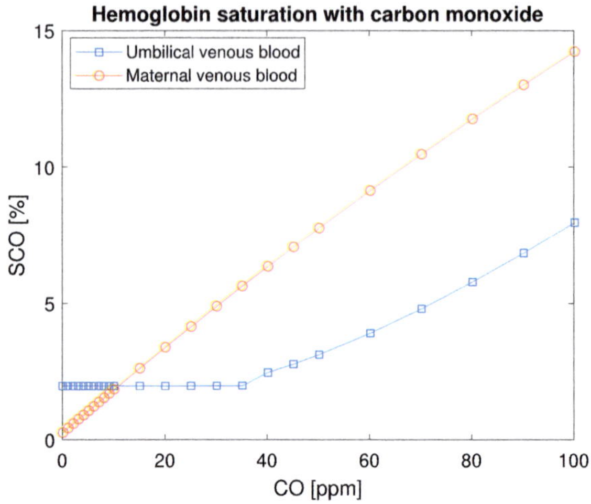

Fig. 7.17 Hemoglobin saturation with carbon monoxide in maternal and fetal blood as a function of carbon monoxide intoxication in ppm. The conversion of CO in ppm to kg/m^3 is to divide by 0.001. *Source* [7]

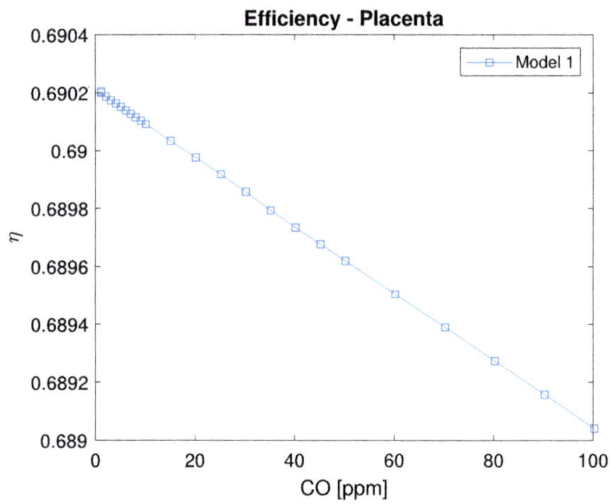

Fig. 7.18 Placenta's exergy efficiency as a function of carbon monoxide intoxication in ppm. The conversion of CO in ppm to kg/m^3 is to divide by 0.001. *Source* [7]

As shown in Fig. 7.18 (model 1), the placenta exergy efficiency is considerably low compared to the respiratory system. This can be explained by its high metabolism, which leads to high irreversibilities, and the fact that the placenta has functions other than providing oxygen to the fetus, which were not taken into consideration in this analysis. This efficiency, calculated by Eq. 6.14, is influenced by the high $\dot{B}_M$ term, which becomes evident by the variation of this physical property as a function of CO in the air, occurring only in the third decimal place.

Figure 7.19 presents the destroyed exergy rate for the placenta. The first and second analyses provide results that differ from those found for the respiratory system. In the first analysis, there is a decrease in efficiency with CO poisoning and an increase in the destroyed exergy rate, while in the second analysis results, there is a reduction in the exergy destruction rate. Only the third analysis results in an increase

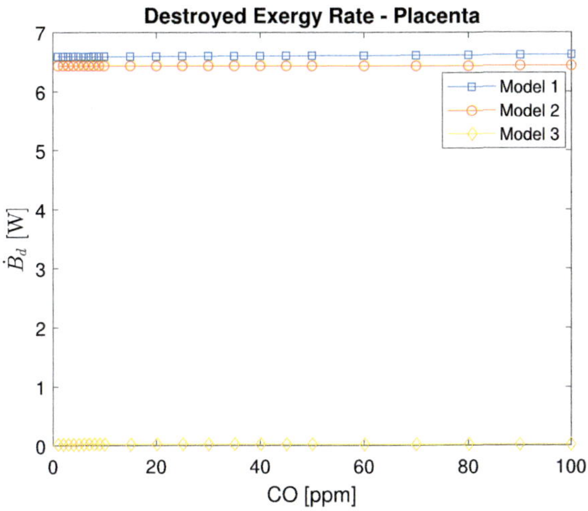

Fig. 7.19 Placenta's exergy destruction rate as a function of carbon monoxide intoxication in ppm. The conversion of CO in ppm to kg/m^3 is to divide by 0.001. *Source* [7]

in the destruction rate for both the respiratory system and the placenta. However, this model yields a destroyed exergy with an order of magnitude lower, indicating that for the placenta, the percentage of irreversibilities associated with mass transfer is not as high as it is for the lungs. This result is significant because both organs are responsible for the exchange of blood with an external environment (for the fetus and mother). This highlights the importance of the placenta, which not only facilitates mass transportation but also supports life. However, it is vital to bear in mind that the irreversibilities increase as a function of carbon monoxide.

The graphic of the destroyed exergy as a function of the percentage of hemoglobin saturated with oxygen in the umbilical venous blood (Fig. 7.20) is similar to

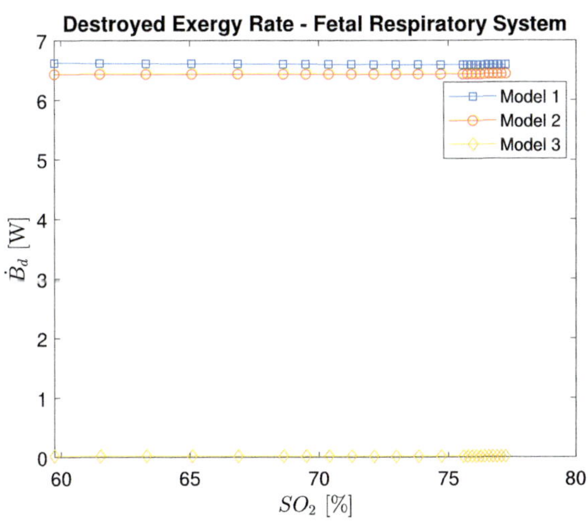

Fig. 7.20 Fetal respiratory system destroyed exergy rate per oxyhemoglobin (saturation of oxygen), which is a function of carbon monoxide poisoning. *Source* [7]

Fig. 7.19. This similarity is due to the fact that the destruction rate is mainly influenced by the metabolism, resulting in relatively constant values. However, it is important to note that the poisoning by CO can still affect the irreversibilities of the placenta, albeit to a lesser extent.

7.6 Concluding Remarks

In this chapter, the method for applying exergy analysis to the human body, initially proposed by [2], was modified to more accurately assess the effects of carbon monoxide poisoning on the exergy behavior of the human respiratory system. Specifically, the focus of the study was on the lungs. The method was applied to the model proposed by [4], allowing for simulations of the respiratory system at different altitudes, with varying levels of acclimatization (changes in hemoglobin concentration and alveolar ventilation), and different levels of carbon monoxide poisoning. Thus, three approaches have been proposed to account for the exergy of the blood and the irreversibilities of mass exchanges. The first approach considered the gases present in the blood (O_2, CO_2, and CO) as ideal mixtures. In the second approach, the binding of O_2 and CO to hemoglobin and the transport of CO_2 in the form of bicarbonate were considered from the perspective of the Second Law of Thermodynamics. Finally, the irreversibilities associated with mass transfer were calculated. Therefore, it is possible to conclude that:

- The first approach leads to an increase in efficiency (and a decrease in the rate of exergy destruction) when carbon monoxide poisoning occurs. It should be emphasized that acclimatization increases exergy efficiency, but as altitude increases, the efficiency decreases.
- For the second model, a significant reduction in the effects of carbon monoxide poisoning can be observed, but it is responsible for an increase in exergy destruction rates. Once again, acclimatization had a positive impact by reducing the exergy destruction rate.
- In the third approach to analyzing irreversibilities associated with mass transfer, there was good agreement with the results of the second evaluation method. This allowed for an increased accuracy in assessing the exergy behavior of the human respiratory system.
- Regarding the mother's exergy behavior of the lungs, it was obtained for the first model, a decrease in the destroyed exergy as a function of the intoxication of CO. The second and third models had similar trends, where the destroyed exergy increases as a function of the CO poisoning. Moreover, the third model indicates the irreversibilities of the mass transfer, which corroborates the results of the second model
- It was also possible to conclude that the main effects in the fetus are caused by hypoxia due to the low oxygen pressures in the fetus's blood. In this analysis, the HbCO concentration was found to be smaller in the fetus's blood than in

the mothers. The exergy efficiency of the maternal respiratory system, calculated from model 1, is higher than the average male. This can be related to increased ventilation of the lungs during pregnancy. Moreover, models 2 and 3 indicated a similar trend which suggests that the main modifications in the lung are associated with the increase of irreversibilities of mass transfer. The placenta destroyed exergy rate is significantly higher than the maternal, due to its high metabolism. The third analysis is useful in this case because it shows that the exergy destroyed rate increases, although in a small amount. This is the reason that makes the other two analysis appears without a slope.

- Eventually, the destroyed exergy rate was correlated with an indicator often used in the medical area, which is the saturation of oxygen in the hemoglobin, showing that the exergy analysis may be a proper index to evaluate the human body in healthy and unhealthy scenarios. This type of study may be incorporated in the medical area first because the exergy analysis may give a complementary diagnosis of diseases. Second, with the phenomenological model, it is possible to suggest different scenarios of intoxication without the necessity of experiments in living beings. A limitation of this study is regarding the saturation curve of the mother subjected to some poisoning, which suffers a change; moreover, the two phenomenological models could be integrated to obtain a more extensive range of results of a pregnant woman.

References

1. Albuqerque C, Pellegrini LF, Ferreira M, Oliveira-Junior S, Yanagihara JI (2010) Exergy analysis of human respiration under physical activity. Int J Thermodyn 13(3):105–109
2. Henriques I, Mady CEK, Albuqerque C, Yanagihara J, Oliveira-Junior S (2014) The effect of altitude and intensity of physical activity on the exergy efficiency of respiratory system. Int J Thermodyn 17(4):265–273
3. Mady CEK, Oliveira-Junior S (2013) Human body exergy metabolism. Int J Thermodyn 16(2):73–80
4. Albuquerque C, Yanagihara JI, Turri F (2008) A carbon monoxide transport model of the human respiratory system applied to urban atmosphere exposure analysis. J Brazilian Soc Mech Sci Eng 30:253–260
5. Albuquerque C, Yanagihara JI (2020) An integrated model of the thermoregulatory and respiratory systems of the human body. Int Commun Heat Mass Transf 116:104683
6. Cenzi JR, Albuquerque C, Mady CEK (2018) The effect of carbon monoxide on the exergy behavior of the lungs. Bioengineering 5(4):108
7. Cenzi JR, Albuquerque C, Mady CEK (2019) Phenomenological and thermodynamic model of gas exchanges in the placenta during pregnancy: a case study of intoxication of carbon monoxide. Int J Environ Res Publ Health 16(21):4138
8. Mountcastle VB (1974) Medical physiology. CV Mosby
9. Szargut J, Morris DR, Steward FR (1988) Exergy analysis of thermal, chemical, and metallurgical processes. Hemisphere Publishing Corporation, New York
10. Dutta A, Chattopadhyay H (2020) Exergetic analysis of human respiratory system including effect of age and gender. Int J Exergy 31(4):370–385
11. Banerjee S, Murmu SC, Chattopadhyay H, Dutta A (2025) Estimation of exergetic behaviour of the human respiratory tract. Int J Exergy 47(1):12–30

12. Fritts HW, Filler J, Fishman AP, Cournand A (1959) The efficiency of ventilation during voluntary hyperpnea: studies in normal subjects and in dyspneic patients with either chronic pulmonary emphysema or obesity. J Clin Invest 38(8):1339–1348
13. DuBois D (1916) A formula to estimate the approximate surface area if height and weight be known. Arch Intern Med 17:863–871
14. Alberty RA, Goldberg RN (1992) Standard thermodynamic formation properties for the adenosine 5'-triphosphate series. Biochemistry 31(43):10610–10615
15. Alberty RA (1998) Calculation of standard transformed Gibbs energies and standard transformed enthalpies of biochemical reactants. Arch Biochem Biophys 353(1):116–130
16. Huang SY, Alexander JK, Grover RF et al (1984) Increased metabolism contributes to increased resting ventilation at high altitude. Resp Physiol 57(3):377–385
17. Ainslie PN, Lucas SJ, Burgess KR (2013) Breathing and sleep at high altitude. Resp Physiol Neurobiol 188(3):233–256
18. Wagman DD, Evans WH, Parker VB, Schumm RH, Halow I (1982) The NBS tables of chemical thermodynamic properties. Selected values for inorganic and C1 and C2 organic substances in SI units. National Standard Reference Data System
19. Perrella M, Sabbioneda L, Samaja M, Rossi-Bernardi L (1986) The intermediate compounds between human hemoglobin and carbon monoxide at equilibrium and during approach to equilibrium. J Biol Chem 261(18):8391–8396
20. De Nevers N, Seader JD (1980) Lost work: A measure of thermodynamic efficiency. Energy 5:757–769
21. Knuttgen HG, Emerson-Junior K (1974) Physiological response to pregnancy at rest and during exercise. J Appl Physiol 36(5):549–553
22. Robson SC, Hunter S, Boys RJ, Dunlop W (1989) Serial study of factors influencing changes in cardiac output during human pregnancy. American J Physiol-Heart Circulatory Physiol 256(4):H1060–H1065
23. Goplerud JM, Delivoria-Papadopoulos M (1985) Physiology of the placenta-gas exchange. Ann Clin Labor Sci 15(4):270–278
24. Acharya G, Sitras V (2009) Oxygen uptake of the human fetus at term. Acta obstetricia et gynecologica scandinavica 88(1):104–109
25. Sharan M, Selvakumar S, Singh MP (1990) Mathematical model for the computation of alveolar partial pressure of carbon monoxide. Int J Biomed Comput 26(3):135–147
26. Guyton AC, Hall JE (2011) Guyton and Hall textbook of medical physiology. Elsevier
27. Aubard Y, Magne I (2000) Carbon monoxide poisoning in pregnancy. British J Obstetrics Gynaecol 107(7):833–838

Chapter 8
Hypothermia as a Therapy

8.1 Introduction

This chapter presents the findings of pertinent papers on literature, including studies by [1–3].

According to [4], hypothermia is characterized by the body reaching temperatures lower than 32 °C. This condition can occur inadvertently: (i) when the body reaches these temperatures due to external conditions (known as deep accidental hypothermia) or (ii) a hypovolemic shock following an accident where the individual experiences blood loss. Another circumstance is when hypothermia can also be intentionally induced as a protective measure after cardiac arrest, aiming for neuroprotective effects [5]. This protection is still being studied in the literature, but some authors relate a decrease in metabolic rate (energy consumption) by 6–7 %/°C, as reported by [2, 3]. For hyperthermia, the opposite is known as the Q_{10} effect as described in [6], where the body may reach hypothermia if submitted to thermal stress (temperatures around 33–35 °C and relative humidity higher than 70 %) with high clothes such as 1 CLO (suit), resulting in an exponential increase.

Based on the severity of the risks associated with lowering body temperature, [7] classifies therapeutic hypothermia into the following categories: (i) mild hypothermia, with body temperature between 34.5 °C and 36.5 °C; (ii) moderate hypothermia, with body temperature between 32 °C and 34.5 °C; (iii) severe hypothermia, with body temperature between 28 °C and 32 °C; and (iv) deep hypothermia, with body temperature below 28 °C.

Karnatovskaia et al. [8] highlight that the therapeutic hypothermia technique was one of the primary lines of treatment to reduce mortality rates in post-arrest patients. However, only a few protocols indicate the best parameters for each anatomy. Moreover, the protocol may vary across different countries and hospitals based on their infrastructure. Some efforts have been made in this direction, as seen in studies such as [9] and [10]. However, some insights into induction parameters have been identified, leading to the best results of neuroprotection, such as after the arrest, a rapid

C. E. Keutenedjian Mady, *Exergy Analysis*, Green Energy and Technology,
https://doi.org/10.1007/978-3-032-01646-1_8

decrease in body temperature (induction stage), followed by maintenance at a temperature between 30 °C and 32 °C (mild hypothermia), avoiding periods exceeding 24 h of care. The rewarming phase (normothermia) should be gradual, with rates of 0.25 °C/h [3]. These conditions result from empirical data, and much is still to be known, as argued by [11]. The positive effects of this treatment have been observed in all three phases [4, 5, 9].

There are various methods available to induce hypothermia; some treatments such as cooling pads and ice baths are applied on the skin; they are called external devices. The other types aim to lower blood temperature using a catheter in the femoral vein, or another large vessel, e.g., internal jugular or subclavian vein. Another approach involves removing the blood and reducing its temperature using an external heat exchanger [12]. Numerous studies have demonstrated the effectiveness of using a catheter [2]. However, other authors have found that in cases where the lean mass is high, reducing skin temperature alone may achieve significant results [3].

The study of hypothermia induction using experiments and computational fluid dynamics has gained increasing attention in recent years, as seen in recent articles published on the topic. However, it is worth noting that research on this subject dates back to the 1970s [13]. Several authors have been studying localized brain hypothermia using an air-insulated catheter [12], and whole-body hypothermia using a catheter in the femoral vein [3]. These studies often compare the results obtained from literature or experimental apparatus with numerical simulation. In addition, the assessment of performance indicators such as effectiveness has become crucial [14]. It is important to highlight that most catheter geometry experiments are designed to avoid invasive applications in living beings [15].

8.2 Methods and Simulations

The circulatory system represents the large vessels using two reservoirs, one of arterial blood and another with venous blood. They exchange heat between themselves but not directly with the tissues, as in Fig. 8.2 in Chap. 3. Pennes' equation [16] describes the enthalpy exchanges between the blood in the small vessels and the tissue. The perfusion rate in each tissue plays a crucial role in characterizing these phenomena ($\mathrm{m}^3_{blood}/\mathrm{m}^3_{tissue}.s = \mathrm{m}^3_{bl}/\mathrm{m}^3_{t}.s$).

8.2.1 Human Thermal Model

Figure 8.1 depicts the human thermal model used in this chapter. The [22] model employs the following parameters for the human body: height 1.76 m, mass 67 kg, body area 1.8 m^2, and volume 0.00627 m^3. To represent the human body in the model, it utilizes 15 cylinders of elliptical cross sections. Additionally, each body segment in the model is composed of a combination of the following tissues: skin, fat, muscle,

Fig. 8.1 Human thermal model used in this chapter to assess thermal comfort conditions, where it is indicated each member and boundary conditions. *Source* [22]

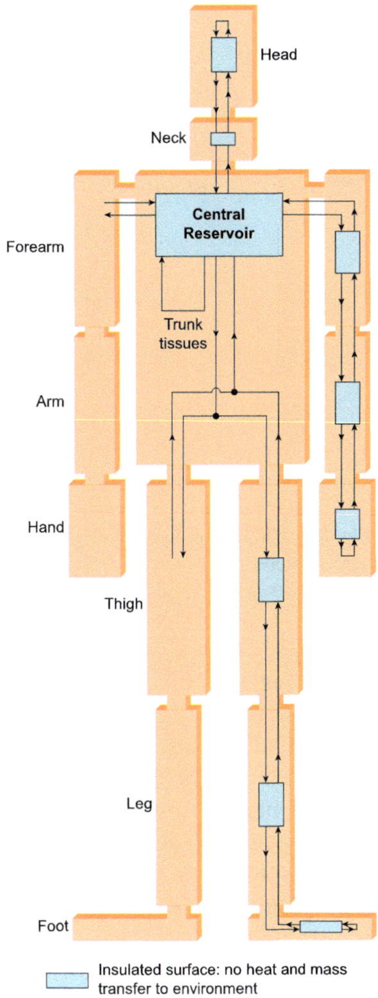

bone, brain, viscera, lung, and heart. The energy balance is solved individually for each tissue, which is assigned its respective metabolism rate, thermophysical properties, and blood perfusion characteristics. The heat exchange between the blood and the tissue is taken into account in the energy balance equation for each element. The vessels are categorized into two types: large and small. The small vessels were treated as part of a continuum, following the model proposed by [16]; later, the model was adjusted and validated by [18]. However, the blood temperature within each tissue is influenced by the location within the model. The treatment of large vessels follows the approach proposed by [19]. In the model, each tissue element is equipped with two blood reservoirs: one for arterial blood and the other for venous blood. These reservoirs facilitate heat exchange between each other, as well as with the adjacent

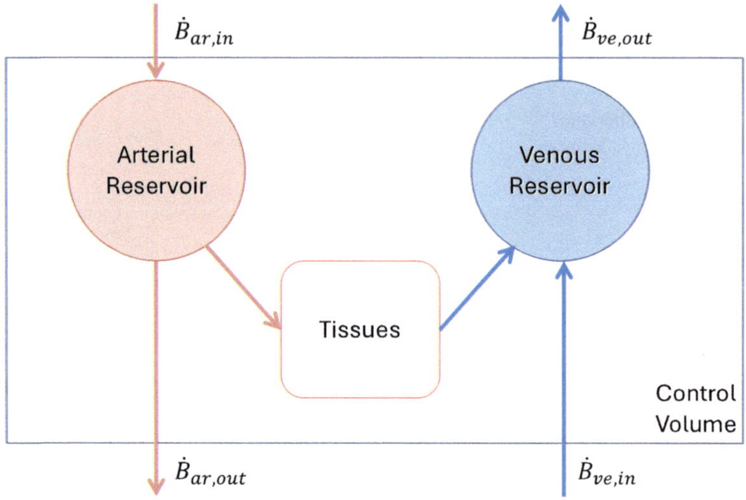

Fig. 8.2 Circulatory system used in each segment. In this figure, it is indicated the arterial and venous blood reservoirs that exchange heat with each other, blood with other compartments and with tissues. *Source* [20, 21] based on [22]

tissues and elements. Notably, the trunk region comprises only one blood reservoir instead of two.

Each arterial reservoir (or a central reservoir in the trunk) represents the large vessels responsible for delivering arterial blood to the tissues, and the venous reservoir receives blood from the small tissue elements, allowing them to reach a state of thermal equilibrium. Considering Eq. 8.1, which represents the enthalpy variation of the blood in the small vessel, where $t = 1, 2, 3, 4$ corresponds to the tissues (skin, fat, muscle, and core), for Fig. 5.1 (simplified model).

$$\Delta \dot{H}_{bl \to n} = \dot{\omega} c_{bl} \rho_{bl} (T_t - T_{bl,out}) \tag{8.1}$$

The consideration of the small vessels as a continuum that exchanges energy with the tissues is derived from Pennes' [16], the revision conducted by Wissler [17], and further validated by Chen [18]. Equation 8.1 is where $\dot{\omega}$ is the blood perfusion rate ($m^3_{bl}/m^3_t.s$), c_{bl} denotes the specific heat of the blood (J/(kg.K)), and ρ_{bl} is the specific mass of the blood (m^3/kg).

For the complete model of Fig. 8.1, the energy balance for the blood can be described by three expressions applied to Fig. 8.2: Eqs. 8.2 and 8.3 for all compartments with two reservoirs; and Eq. 8.4 for the trunk. The first terms of the right-hand side of Eqs. 8.2 and 8.3 refer to the enthalpy entering the control volume due to blood coming from other elements. The integral that is appearing in Eq. 8.4 refers to the enthalpy entering the venous reservoir from the blood in the small vessels, which leaves the tissues in thermal equilibrium, i.e., with temperature T_t (where t represents the skin, fat, and muscle, among others). In the trunk, only one blood reservoir was considered. Thus, there is no heat transfer between arteries and veins, leaving only the terms related to the enthalpies entering and leaving the reservoir. The first term in Eq. 8.4 refers to the enthalpy of venous blood coming from the different compartments i of the body.

$$m_{ar,i}c_{bl}\frac{dT_{ar,i}}{dt} = \rho_{bl}c_{bl}\dot{V}_{ar,i}\left(T^{e}_{ar,i} - T_{ar,i}\right) + h_{av}\left(T_{ve,i} - T_{ar,i}\right) \tag{8.2}$$

$$m_{ve,i}c_{bl}\frac{dT_{ve,i}}{dt} = \rho_{bl}c_{bl}\dot{V}_{ve,i}\left(T^{e}_{ve,i} - T_{ve,i}\right) + h_{av}\left(T_{ar,i} - T_{ve,i}\right) + \int_{V}\rho_{bl}c_{bl}\dot{\omega}_{bl}\left(T_{t} - T_{ve,i}\right)dV \tag{8.3}$$

$$m_{s}c_{bl}\frac{dT_{s,i}}{dt} = \sum_{i}\rho_{bl}c_{bl}\dot{V}_{ve,i}\left(T_{ve,i} - T_{bl}\right) + \int_{V}\rho_{bl}c_{bl}\dot{\omega}_{bl}\left(T_{t} - T_{bl}\right)dV. \tag{8.4}$$

In which: $m_{ar,i}$, $m_{ve,i}$, and m_{bl}: the masses of the arterial, venous, and trunk reservoirs; c_{bl}: the specific heat of blood; ρ_{bl}: the specific mass of blood; $\dot{V}$: the arterial (subscript ar) or venous (subscript ve) blood flow rate; $T^{e}_{ar,i}$, $T^{e}_{ve,i}$: the temperatures of arterial and venous blood entering the control volume; $T_{ar,i}$, $T_{ve,i}$, T_{bl}: the temperatures of the arterial, venous, and central reservoirs; h_{av}: the heat transfer coefficient between the large arteries and veins; T_{t}: the temperature of the blood coming from the tissues; and $\dot{\omega}_{bl}$: blood flow in the tissues.

8.2.2 *Model Modification for Hypothermia*

According to [22], having an initial condition is required for solving the differential equations for the transient simulations. To do so, the thermal neutrality condition must be determined, where the temperatures of the tissues and blood in different compartments are at equilibrium. These temperatures can then be used as the initial condition for the transient simulations. According to [24], operative temperatures between 29 and 31 °C result in the thermal neutrality condition of the body with no clothes. This condition is when all thermoregulatory control mechanisms of the body, such as sweating and shivering, are not active. More information on the energy equation solved for each tissue and member is in [22].

The exergy analysis is modified with one additional term for the extracorporeal blood removal (Eq. 8.5). Additionally, there is heat removal by a catheter or cooling pads as in Eq. 8.6.

$$\dot{B}_{dest} = \dot{B}_{M} - \left.\frac{d\mathbf{B}_{corpo}}{dt}\right|_{\Delta T} - \left(\dot{B}_{r} + \dot{B}_{c} + \Delta\dot{B}_{res} + \dot{B}_{e}\right) - \dot{W} + \left(\dot{B}_{hyp} - \dot{B}_{bl}\right) \tag{8.5}$$

$$\dot{B}_{dest} = \dot{B}_{M} - \left.\frac{d\mathbf{B}_{corpo}}{dt}\right|_{\Delta T} - \left(\dot{B}_{r} + \dot{B}_{c} + \Delta\dot{B}_{res} + \dot{B}_{e}\right) - \dot{W} + \dot{B}_{Q_{hyp}} \tag{8.6}$$

The initial results of the group were obtained by [2] for three types of tests: internal hypothermia using a catheter with constant value of UA (W/m^2), external cooling of the blood itself, and attempts to induce hypothermia using the model proposed by [22]. The geometry of the catheter with a saline balloon heat exchanger is complex, and the blood flow characteristics are not well defined. Therefore, based on the experimental results of [23], the product of the heat transfer coefficient of the catheter and the surface area (UA) was determined to be 62 W/K (Fig. 8.3).

In the first two techniques, a catheter is inserted into the vena cava through the femoral vein. However, in the second procedure, only one catheter with a triple-lobed, helically wound, heat exchanger balloon is inserted. In contrast, the first technique requires the use of two catheters, one to remove the blood and another to reintroduce it. Both techniques may be employed to induce either hyperthermia or hypothermia. In all procedures, the model proposed by [22] had to be adapted for the simulations. The circulatory system was divided into small vessels (treated as a part of continuum with the tissues) and large vessels (modeled as two reservoirs, one for arterial blood and the other for venous blood). The trunk was modeled with only one reservoir, where venous blood flows into. Figures 8.4 and 8.5 depict the catheter employed in the simulation, which is crucial to highlight that the dimensions are based on the manufacturer's catalog of Zoll Cool Line©. The catheter has an outer diameter equivalent to gauge 9.3 F (3.1 mm), and a length of 27 cm. According to the findings by [15], a four-balloon array provides improved cooling effects compared to a lower number of balloons. Using more balloons could lead to temperature saturation across different positions. Based on this information, we opted for a four-balloon pattern. The wall thickness is set at 0.1 mm.

Figures 8.6 and 8.7 present a comparison of all techniques studied for inducing hypothermia. Figure 8.6a, b shows the temperature of the hypothalamus as a function of time. The destroyed exergy and exergy efficiencies as a function of time are displayed in Figure 8.7a, b, respectively. To address the challenge of incorporating two environmental parameters in the same simulation, the water bath technique was analyzed separately during the cooling period. However, Mady et al. [2] show that

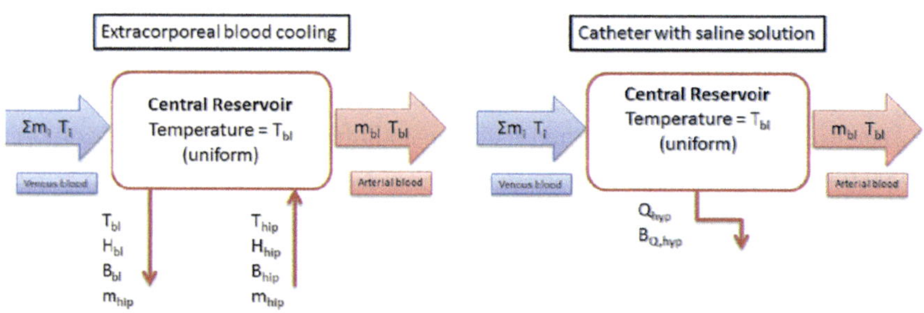

Fig. 8.3 Adaptation of the central reservoir of the model proposed by [22] for two of the three procedures (1) Extracorporeal blood cooling (EBC). (2) Endovascular cooling with catheter (ECC). (3) Water bath (WB). *Source* [2]

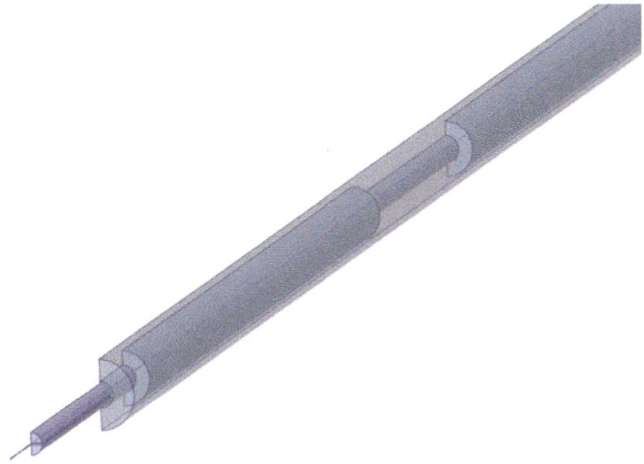

Fig. 8.4 A three-dimensional model of a cooling catheter. Outer view of the catheter. The external cylinder would be a simple representation of the veins. *Source* [1]

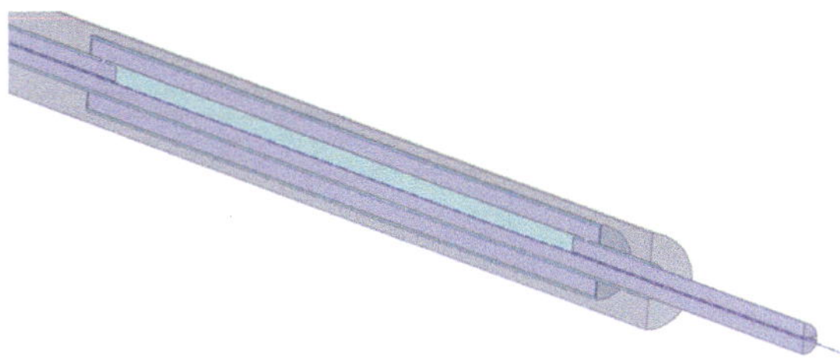

Fig. 8.5 A three-dimensional model of a cooling catheter. Longitudinal section view of the saline infusion trajectory. *Source* [1]

this specific technique was found to be less effective compared to other techniques in the anatomy of [22], consistent with the findings of [26]. This technique resulted in a higher hypothalamus temperature for the same target temperature, as shown in Fig. 8.6a. External cooling operates by increasing the energy transfer from the skin to a specific device in the environment, which opposes the passive and active systems of the human body, as depicted in Fig. 8.6b. Therefore, the body increases the energy released through mechanisms like shivering to compensate for the heat loss, making it necessary to inhibit the body's natural control system to decrease the core temperature effectively.

However, Fig. 8.6b suggests that the passive and active systems are adequate in responding to these external "stresses" (indicated in black). However, the passive system alone is insufficient to inhibit hypothermia (indicated in blue). When the body's control system is active, the model does not enter into hypothermia. On the other hand, when the control system is inactive, the water bath technique successfully induces the body into this state. If the test is not applied directly to the skin, therefore internal, the passive and active systems are inefficient in preventing hypothermia.

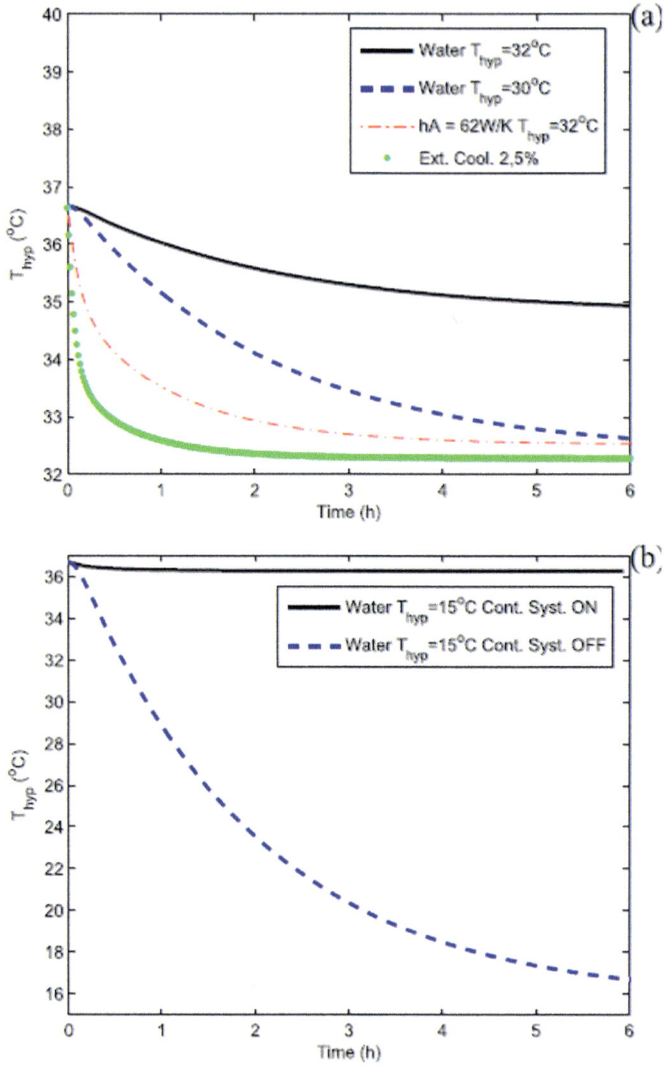

Fig. 8.6 Hypothalamic temperature for: **a** different techniques of hypothermia **b** water bath with active and inactive control system. *Source* [2]

This highlights the advantages of the other techniques compared to water baths. Further explanations regarding these results are in the study conducted by [3].

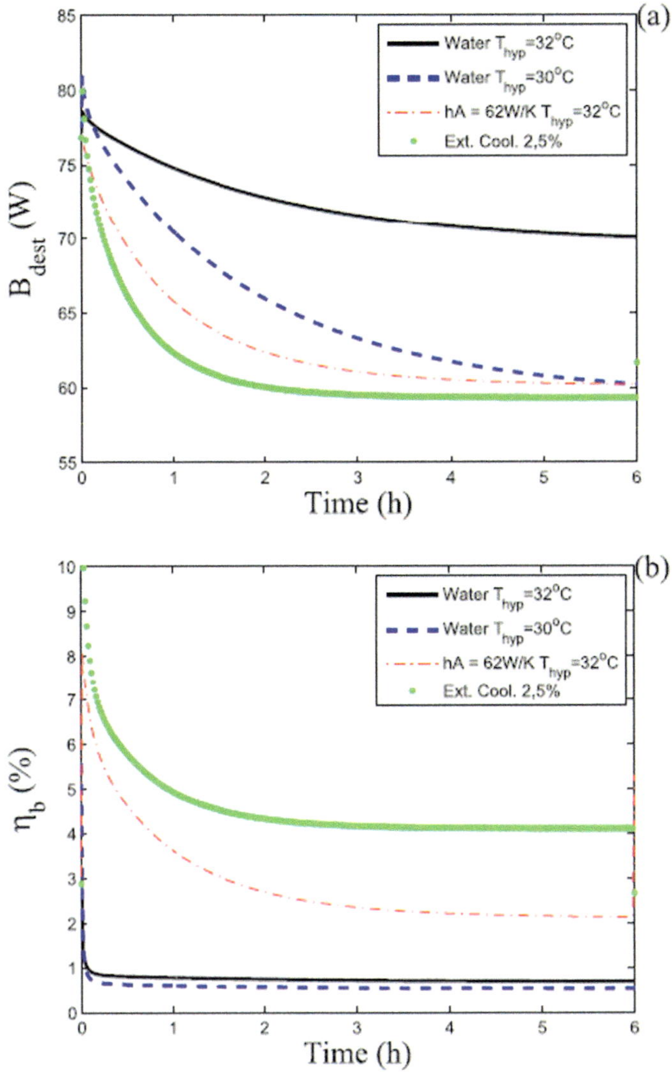

Fig. 8.7 Destroyed exergy rate ($\dot{B}_{dest}$) and exergy efficiency (η_b) as a function of time for three types of cooling techniques. *Source* [2]

8.3 Results and Discussions

Based on the findings depicted in Figs. 8.6 and 8.7, it can be concluded that the extracorporeal blood cooling technique resulted in a hypothalamus temperature closest to 32 °C. Additionally, it exhibited lower values of $\dot{B}_{dest}$ and higher values of η_b. Figure 8.7b suggests that the values of hb are smaller for extracorporeal blood cooling due to a smaller Carnot factor between the hypothalamic and environmental temperature. Hence, by considering the principles of the Second Law of Thermodynamics, it becomes feasible to assess the quality of the energy interactions that occur in the tests and the body's performance throughout the tests.

In light of these findings, Alves and Mady [3] proposed three models incorporating combinations of external and internal cooling techniques involving catheters and cooling pads. The authors considered that currently there is no protocol for simulations, and some hospitals do not have a specific protocol in place at all. It is worth noting that the majority of the results are derived from animal or clinical experiments. One limitation encountered was the fixed geometry of the model proposed by [22], which made it challenging to change fat and muscle masses computationally. That would mischaracterize the model, and it would be necessary to revalidate the outcomes of thermal behavior. Figure 8.8 illustrates three models for inducing hypothermia, the first being external cooling using cooling pads, while the second model utilizes a catheter and a combination of both.

The treatment environment is typically maintained at a temperature of 24 °C and relative humidity of 50% [24]. For these situations, all the models start from the thermal neutrality condition [24], which serves as the initial value for solving the differential equations of the model proposed by Ferreira and Yanagihara [22]. Each simulation starts at the same temperature distribution and is then subjected

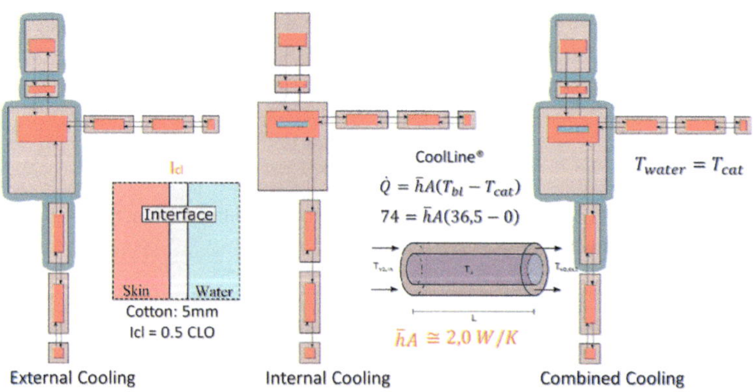

Fig. 8.8 Adaptation of the central reservoir of the model proposed by [22] for two of the three procedures (1) Extracorporeal blood cooling (EBC). (2) Endovascular cooling with catheter (ECC). (3) Water bath (WB). Considering that T_{air} is 24 °C and ϕ of 50%. *Source* [3]

Table 8.1 Device temperature for each phase of the simulations. *Source* [3]

Method	External °C	Internal °C	Combined °C
Induction	4	4	4
Maintenance	31	40	31
Rewarming	40	40	40

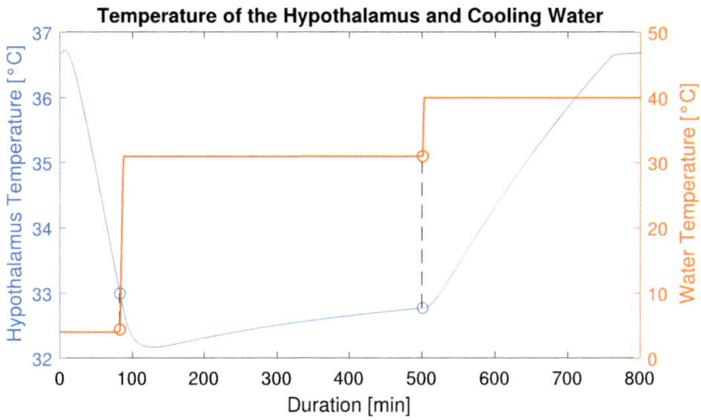

Fig. 8.9 Example of hypothalamus and device temperatures during simulations. *Source* [3]

to hypothermia induction employing an external, internal, or combined cooling device [3]. The temperature of each device is adjusted based on the hypothalamus temperature (refer to Table 8.1 and Fig. 8.9).

At the beginning, each device is set to the lowest acceptable value of 4 °C in order to evaluate its heat removal capability. As the hypothalamus temperature reaches 33 °C (which is the limit of mild hypothermia [25]), the device maintains the temperature specified in Table 8.1. During the rewarming phase, the devices are heated gradually until they reach their maximum acceptable temperature of 40 °C. This allows the measurement of their maximum rewarming rate. The temperature of each device varies at a rate of 0.1 °C/s during the transition period.

External cooling tools can range from simple water-soaked towels to arrangements of cooling pads. In this study, the arrangement was chosen to represent the external cooling tools due to its ability to effectively control the patient's core temperature. By using cooling pads in the model, the model represents the usual placement of these devices, exchanging heat only with the pads, while the rest of the body is exposed to the usual heat transfer phenomena as described in [22]. In order to simulate the interface material, the heat resistance of a thin layer of cotton was calculated using McCullough and Jones' equation, as described in [24], resulting in a value of 0.2 CLO (or 0.031 m^2K/W).

The catheter is a commonly used tool for controlling temperature in patients [1–3]. It is a tube inserted through the femoral, internal jugular, or subclavian veins, with balloons that inflate once it reaches the superior or inferior vena cava, circulating a saline solution at the desired temperature in its interior. In most cases, catheters are designed for specific insertion sites due to differences in blood vessel diameters (an exception is the Cool Line catheter produced by Zoll Medical Corp., US). In this apparatus, a certain amount of heat flow rate ($\dot{Q}_{cat}$) is removed from the central blood reservoir (a location that best represents the vena cava [2]), calculated using Newton's law of cooling: $\dot{Q}_{cat} = hA_{cat}(T_{cat} - T_{bl})$. While the temperatures of the blood (T_{bl}) and the catheter (T_{cat}) are determined at each time step, the product hA remains constant. By referring to the datasheet for the product, the values can be estimated to be approximately 2.0 W/K. Similarly, Eq. 8.7 estimates the exergy transported with $\dot{Q}_{cat}$, by considering the logarithmic mean temperature difference between the blood in the previous time step and after the energy balance (T_{ml}), indicated in Eq. 8.8.

$$\dot{B}_{cat} = \dot{Q}_{cat}\left(1 - \frac{T_0}{T_{ml}}\right) \tag{8.7}$$

$$T_{ml} = \frac{\left(T_{bl}^{i} - T_{bl}^{i-1}\right)}{ln\left(\frac{T_{bl}^{i}}{T_{bl}^{i-1}}\right)} \tag{8.8}$$

For the assessment of the efficacy and risks of each method, Figs. 8.10, 8.11, and 8.12 provide comparisons of the temperatures for the brain, heart, visceral organs, and mean skeletal muscles during the different phases of therapy for external, internal, and combined cooling devices. Figure 8.10a illustrates that in the external device model, mild hypothermia is achieved after 65 min of cooling. Once the water flow is warmed, it takes approximately 40 min for the vital organs to reach a stable condition during both the maintenance and rewarming stages. One possible explanation is the method itself [2]. As the heat is removed through the skin, it relies on the conduction of heat through the surrounding tissue that encompasses the internal organs and major blood vessels. The delay observed between alterations in water flow and tissue can be attributed to the insulating properties of fat. Therefore, the higher the body mass of the patient, the higher the expected delay. Figure 8.10b illustrates that, although changes in the catheter's condition can be observed in less than 10 min, the device itself requires 79 min to lower the brain's temperature below 34 °C and was incapable of maintaining this temperature, even when operating at its maximum heat transfer capability (surface temperature at 40 °C). Lastly, Fig. 8.10c shows that the combined method has a similar response, with the difference being the response time. The time required to cool the brain to 34 °C was reduced to 40 min, and stability was achieved 20 min after the changes in the device's condition.

The heat removed by the catheter represents a higher quality heat transfer compared to the heat removed at the skin surface, leading to a more pronounced impact on the body's exergy profile. Additionally, based on the response of the combined method, it can be assumed that each tool does not interfere with the other, considering

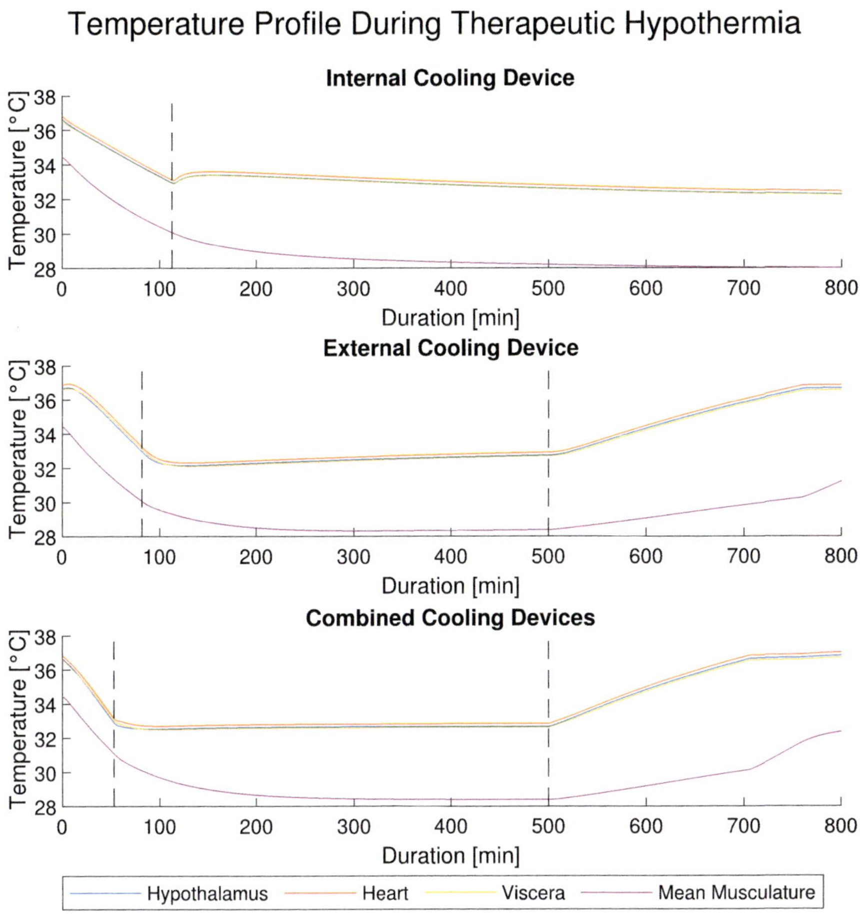

Fig. 8.10 Temperature profile of different tissues over time for each cooling method. *Source* [3]

that the internal technique has a greater influence during temperature changes, while the external technique comes into play after an adjustment period of 20 min.

Figure 8.12 illustrates the rate at which the exergy is destroyed. All models exhibit a similar trend as the body's core temperature during each phase, although there are variations in absolute values. This observation highlights that, while the human body has an optimal temperature for its physiological processes, deviating from this temperature through hypothermia does not improve their quality, but significantly slows down their rate, resulting in a reduced rate of exergy destruction. The variations observed in Fig. 8.12 can be attributed to the temperature profiles and exergy flows associated with the medical equipment. The exergy flows to the environment, in comparison, are negligible. Compared to the external model, the catheter decreases

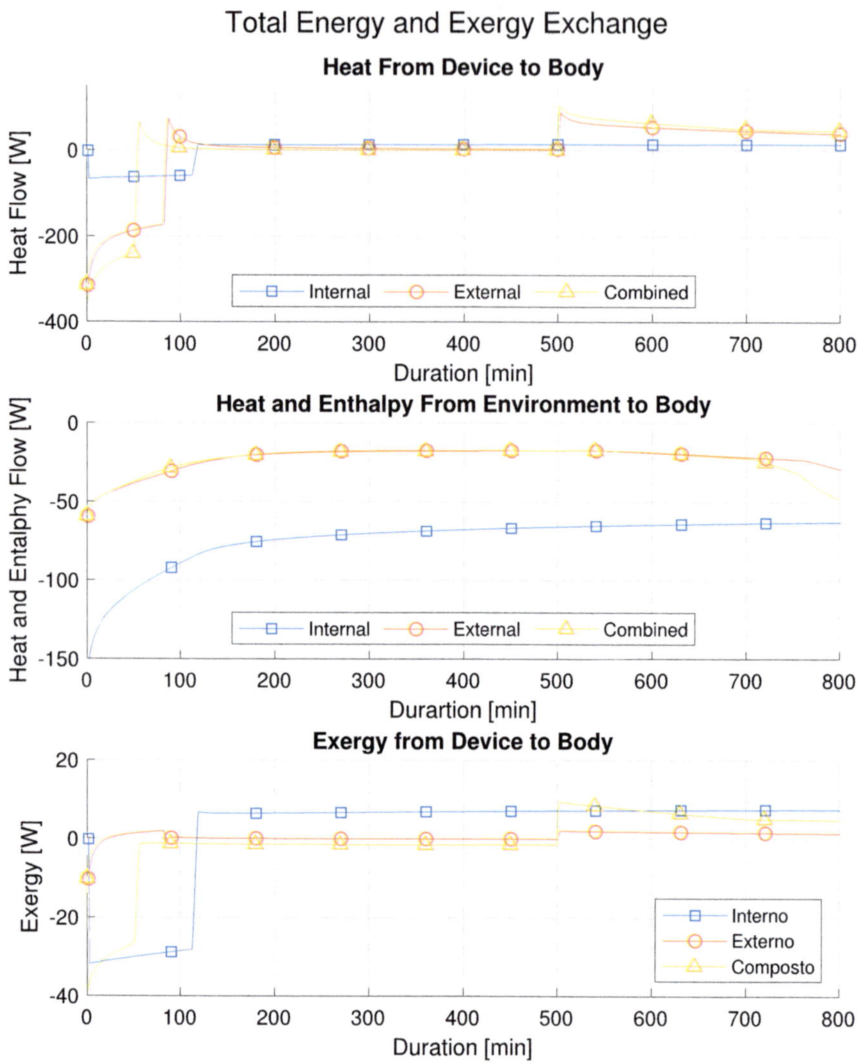

Fig. 8.11 Energy transfer with catheter **a** and cooling pads **b** for each method. *Source* [3]

the irreversibilities during the induction period but increases them during maintenance and rewarming. The increase in irreversibilities is more pronounced in the internal model compared to the combined model. This is because the combined model achieves stability with a lower heat flow, resulting in a reduction of the exergy inserted by the catheter.

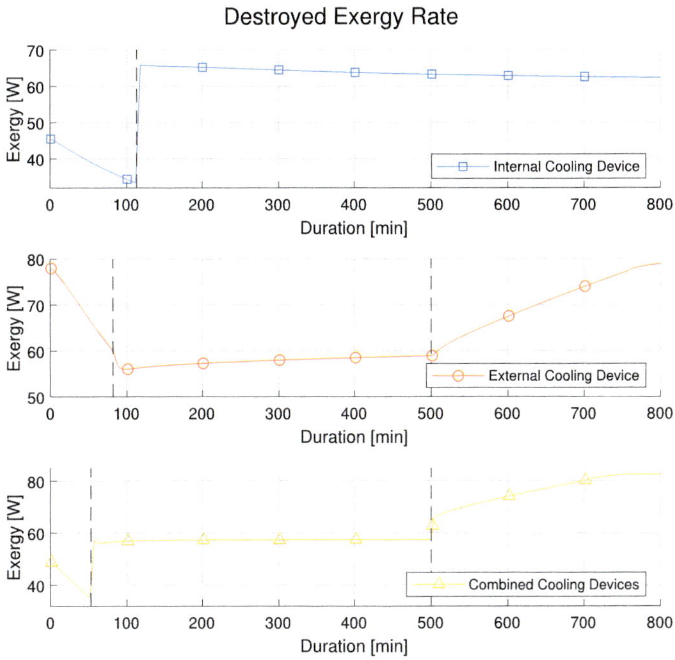

Fig. 8.12 Destroyed exergy rate for each method. **a** Catheter, **b** cooling pads, and **c** combined method. *Source* [3]

8.4 Concluding Remarks

The human thermal model developed by [22] allowed for the comparison of various hypothermia techniques described in medical literature and the analysis of different parameters of each cooling technique from the perspectives of energy and exergy analyses without using live subjects (animals and humans).

The findings of [2] allow for a proposal of suitable hypothermia protocols in hospitals, considering the exergy analysis. These analyses serve as a prominent feature of the chapter and highlight the significance of the control system on the outcomes of each technique. A comparison of the skin water bath, the catheter with saline solution, and the extracorporeal blood cooling techniques showed that the extracorporeal blood cooling was most effective (from the exergy point of view) and had lower destroyed exergy. Furthermore, during the procedure, the model reached the target temperature faster with higher rewarming rates. In the case of the skin water bath technique, the control mechanism prevented the body from entering a state of hypothermia, making it a complementary technique or requiring the use of a mechanism to inhibit shivering as proposed by [3]. However, for the other two techniques, the shivering mechanism had a negligible effect, with a temperature difference of no lower than 1% when the mechanism was turned on or off.

In all categories, the combined method in the study conducted by [3] showed better results due to the absence of interference from the combined effects of each mechanism. In other words, the catheter provides faster and more localized cooling, reducing irreversibilities during the induction phase, while the cooling blanket maintains a higher heat removal capacity and lower exergy destruction during the maintenance and rewarming phases. If applied individually, internal cooling has a faster effect on the core temperature variation, but can be influenced by heat loss to the environment (especially on the anatomy of the model of [22]. In contrast, external cooling has a greater capacity to modify the core temperature, but its effectiveness may vary depending on the patient's body type. Thus, in cases where the patient has a higher body fat percentage, internal cooling may be more suitable. On the other hand, external cooling is generally preferred for patients with low body fat. In future research, the internal model can be refined by reducing the exposed area, such as by utilizing isolating blankets. Likewise, the combined model could employ different elements or techniques during different phases. For example, a catheter could be used solely during the induction phase and initial maintenance period. Additionally, the model currently utilized can still be improved, becoming capable of simulating both the injuries and the healing process in order to assess their influence on the exergy destruction.

References

1. Cavalcanti GM (2024) Análise numérica de um cateter de resfriamento para indução de hipotermia em pacientes pós parada cardiorrespiratória. Master dissertation University Center of FEI. https://doi.org/10.31414/EM.2024.D.131764
2. Mady CEK, Henriques IB, Oliveira-Junior S (2015) A thermodynamic assessment of therapeutic hypothermia techniques. Energy 85:392–402
3. Alves EC, Mady CEK (2020) Thermodynamic assessment of therapeutic hypothermia techniques for rehabilitation of post-cardiac arrest patients. Case Stud Thermal Eng 22:100752
4. Swol J et al (2022) Extracorporeal life support in accidental hypothermia with cardiac arrest-a narrative review. Asaio J 68(2):153–162
5. Presciutti A, Perman SM (2022) The evolution of hypothermia for neuroprotection after cardiac arrest: a history in the making. Ann New York Acad Sci 1507(1):60–69
6. Mady CEK, Albuquerque C, Fernandes TL, Hernandez AJ, Yanagihara JI, Oliveira-Junior S (2014) Human body exergy analysis and the assessment of thermal comfort conditions. Int J Heat Mass Transf 77:577–584
7. Andresen M, Gazmuri JT, Marín A, Regueira T, Rovegno M (2015) Therapeutic hypothermia for acute brain injuries. Scandinavian J Trauma Resuscitation Emer Med 23:1–7
8. Karnatovskaia LV, Wartenberg KE, Freeman WD (2014) Therapeutic hypothermia for neuroprotection: history, mechanisms, risks, and clinical applications. Neurohospitalist 4(3):153–163
9. Nolan JP et al (2008) Post-cardiac arrest syndrome: epidemiology, pathophysiology, treatment, and prognostication: a scientific statement from the International liaison committee on resuscitation; the American heart Association emergency cardiovascular care Committee; the Council on cardiovascular surgery and anesthesia; the Council on cardiopulmonary, perioperative, and critical care; the Council on clinical cardiology; the Council on stroke. Resuscitation 79(3):350–379
10. Orban JC et al (2012) The practice of therapeutic hypothermia after cardiac arrest in France: a national survey. PloS one 7(9):1–5

11. Wu D et al (2020) Selective intra-arterial brain cooling improves long-term outcomes in a non-human primate model of embolic stroke: efficacy depending on reperfusion status. J Cerebral Blood Flow Metabol 40(7):1415–1426
12. Jiang M et al (2022) The blood heat exchanger in intra-arterial selective cooling infusion for acute ischemic stroke: A computational fluid-thermodynamics performance, experimental assessment and evaluation on the brain temperature. Comput Biol Med 145:105497
13. Ahuja AS, Hendee WR (1978) Thermal design of a heat exchanger for heating or cooling blood. Phys Med Biol 23(5):937
14. Merrill TL, Mitchell JE, Merrill DR (2016) Heat transfer analysis of catheters used for localized tissue cooling to attenuate reperfusion injury. Med Eng Phys 38(8):758–766
15. Cattaneo G, Schumacher M, Wolfertz J, Jost T, Meckel S (2015) Combined selective cerebral hypothermia and mechanical artery recanalization in acute ischemic stroke: in vitro study of cooling performance. American J Neuroradiol 36(11):2114–2120
16. Pennes HH (1948) Analysis of tissue and arterial blood temperatures in the resting human forearm. J Appl Physiol 1(2):93–122
17. Wissler EH (1998) Pennes' 1948 paper revisited. J Appl Physiol 85(1):35–41
18. Chen MM, Holmes KR (1980) Microvascular contributions in tissue heat transfer. Ann New York Acad Sci 335(1):137–150
19. Wissler EH (1985) Mathematical simulation of human thermal behavior using whole body models. Heat Transf Med Biol 1(13):325–373
20. Mady CEK, Ferreira MS, Yanagihara JI, Saldiva PHN, de Oliveira Junior S (2012) Modeling the exergy behavior of human body. Energy 45(1):546–553
21. Silva-Ribeiro TJ, Mady CEK (2022) Comparison among exergy analysis methods applied to a human body thermal model. Energy 239:122446
22. Ferreira MS, Yanagihara JI (2009) A transient three-dimensional heat transfer model of the human body. Int Commun Heat Mass Transf 36(7):718–724
23. Dixon SR et al (2002) Induction of mild systemic hypothermia with endovascular cooling during primary percutaneous coronary intervention for acute myocardial infarction. J American College Cardiol 40(11):1928–1934
24. American Society of Heating, Refrigerating, Air-Conditioning Engineers (2005) ASHRAE handbook
25. Peng RY, Bongard FS (1999) Hypothermia in trauma patients. J American College Surg 188(6):685–696
26. Finley AC, Rachabattula S, Eyngorn I et al (2011) A comparison of cooling techniques to treat cardiac arrest patients with hypothermia. Stroke Res Treat 1:690506

Chapter 9
Exergy Analysis of Brazilian Society

9.1 Overall Exergy Analysis of a Society

Exergy analysis was conducted in the state of São Paulo, Brazil, to assess exergy efficiencies of its constituent sectors, namely, Industrial, Rural, Commercial and Public, Residential, and Transportation. The sectors were ranked from most to least efficient. The overall exergy efficiency was determined to be 28.3%, a value which falls within the range of similar studies. The control volume used in the analysis is in Fig. 1.2 described in Chap. 1. The pioneering study on exergy analysis of society was conducted by Reistad [1], where he calculated an exergy efficiency of 21% for the United States. In this innovative approach, society was divided into different sectors to provide a better characterization of end-use efficiencies. It focused solely on energy carrier flows, while any remaining exergy that was destroyed in the environment was considered external irreversibilities. This text followed the methodology proposed by [1]. Another approach was introduced by Wall [2, 3] where the author considered energy carriers for non-energy use as well, such as raw materials for construction or industrial processes, similar to the extended exergy analysis proposed by [4]. This method is widely used in further analysis when the extended exergy analysis is used in combination with life-cycle analysis. These three studies reported exergy efficiencies of 25%, 18%, and 22%, respectively.

Since these pioneering works, other studies have been conducted for various societies, adopting one of the previously mentioned approaches. Utlu and Hepbasli [5] assessed the exergy efficiency of 24.78% for the residential and commercial sectors in Turkey. Ertesvag and Mielnik [6] determined exergy efficiencies of 21% and 15% for the base years 1985 and 1990, respectively. In subsequent work, Ertesvag [7] compared the results obtained for different societies and found similarities. In Brazil, a similar study was conducted by Schaeffer and Wirtshafter [8] for the base year of 1987. The study revealed an overall exergy efficiency for the country of 23%, highlighting the disparity between production and end-use quality and emphasizing the importance of rational exergy utilization. Rosen [9] compared the overall energy

C. E. Keutenedjian Mady, *Exergy Analysis*, Green Energy and Technology,
https://doi.org/10.1007/978-3-032-01646-1_9

efficiency, which was found to be 50%, and the lower exergy efficiency of 24%. The study revealed a significant disparity between these indicators, particularly in Canada's Residential, Commercial, and Industrial sectors. In a later study by [10], the overall exergy efficiency for São Paulo state was found to be 28.3%, which falls within the range of values reported in the literature. The Industrial Sector exhibited the highest efficiency (45.5%), followed by the rural sector, with (31.7%). For the commercial and public, residential, and transportation sectors, exergy efficiencies were lower, with values of 19.2%, 12.4%, and 10.6%, respectively. Therefore, it is essential to assess these indexes in Brazil and [11] calculated that the residential sector has an exergy efficiency of 13%. Eventually, this type of analysis has garnered significant attention in the literature, prompting the evaluation of these figures in other countries. For example, studies were conducted in countries like China [12] and Bangladesh [13].

Therefore, the results of exergy efficiency obtained in these studies ranged from 10 to 30%. Each study may have different assumptions and modeling decisions. There is currently no unified guideline approach for performing an exergy analysis for a society. Nevertheless, they share the same general steps, namely: (i) proceeding from official statistics regarding energy use, (ii) applying the appropriate process parameters, (iii) applying the exergy concept, and (iv) a final exergy efficiency index. An attempt has yet to investigate whether population, environmental conditions, financial conditions, or other variables influence efficiency, and the reason mentioned above would impose significant barriers.

Nevertheless, according to Mosquim et al. [10] society still needs to be aware of this finding and strive to improve the quality of energy conversion processes as a collective effort. The results highlight, for instance, that the Brazilian electrical mix demonstrates high efficiency. However, considering the size of the country's economy, there is room for improvement in the transportation sector, with a lower carbon footprint. Figure 9.1 shows a photo of one state in Brazil which is quite different from other states due to an inequality of the economic, social, climate, and other parameters in a continental country.

Promoting the rational use of exergy, therefore, is a matter of great importance for society and aligns with the Sustainable Development Goals (SDGs) set by the United Nations. The industrial sector has investments aimed at improving the quality of the energy conversion process. Works such as Mady et al. [14] and Lima and Mady [15] that are in the best interest of each plant to improve the efficiency have a direct increase in the outcomes and profits.

Figures such as the exergy efficiency of the transportation sector, which stands at 10.6% (947.6 PJ of exergy consumption), and the residential sector, at 12.4% (198.7 PJ of exergy), are striking to anyone concerned about investing in public policies. Is it necessary to transition from ICE vehicles to electric vehicles? How to decarbonize the transportation sector? How to improve energy efficiency in residences? Is it possible to change people's habits? The following sections discuss these questions, considering that only the passenger fleet was studied, and further research is needed to address the entire transportation sector. One of the conclusions from the study conducted by [10] is that exergy efficiency has limitations when applied

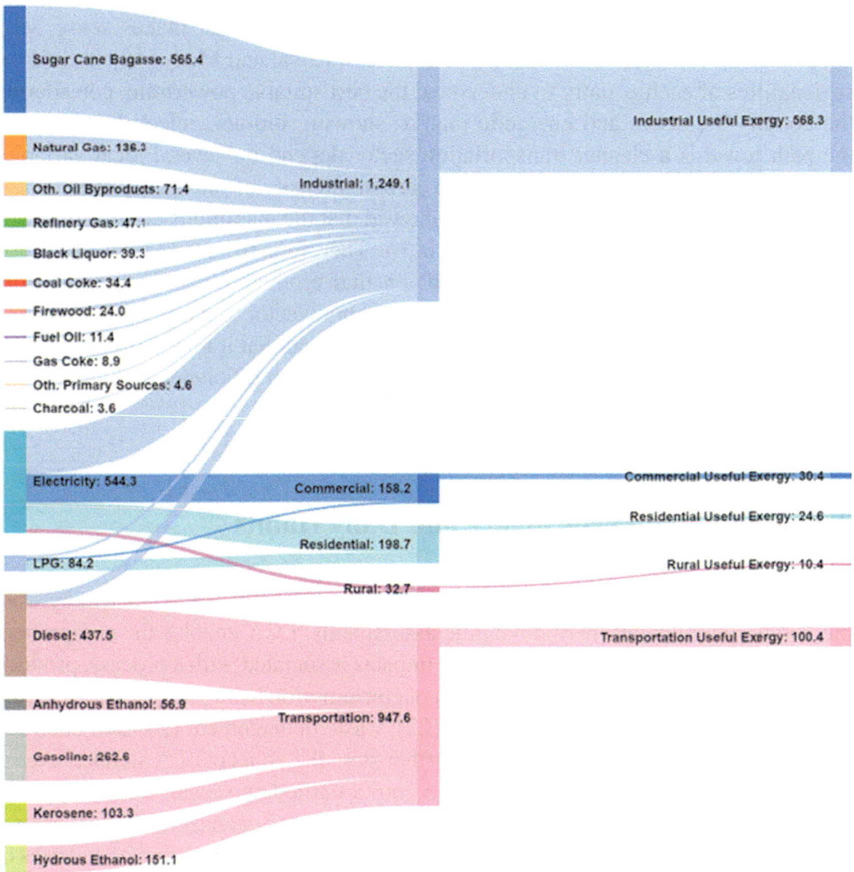

Fig. 9.1 Exergy flows in the state of São Paulo (in PJ). Mosquim et al. [10]

to the transportation sector, particularly due to similar efficiencies observed across different fuel types. Nevertheless, this discussion brings light to extended exergy analyses [4] and its complementary evaluation with life-cycle exergy analysis to this sector [16], Rovai et al. [16] evaluated the potential avoided emissions; specifically, the carbon footprint was used as the metric in the study. The results indicated that a fleet consisting of 90% flex-fuel cars running on ethanol could potentially reduce emissions by 26.8% in an optimistic scenario. On the other hand, fleet electrification emissions were found to be above baseline levels, ranging from 1% to as much as 37%, depending on the energy generation technology and the carbon footprint of the battery.

Feliciano et al. [17] obtained that the exergy efficiency of the electric vehicle in Brazil is higher than using conventional internal combustion engines, taking into consideration from the wheel to the well only of the fuel. One interesting point is considering the electrical mix of Brazil; this fact is actual, but a small change to a more carbon-intensive mix switches to the scenario that makes sense, with ethanol being less carbon-intensive. Nevertheless, Rovai and Mady [18] studied the regionalities of each country to understand the best suitable powertrain, considering the country's electric and energetic matrix, showing biofuels, electrification, and the path towards a cleaner transportation sector depend on several local variables including the availability of land and the carbon intensity of the energy mix.

Finally, Righetto, and Mady [20, 21] showed that the maximum exergy efficiency of photosynthesis is 10% and the average is 5%. Therefore, from the exergy efficiency point of view, producing ethanol is less efficient than producing electrical energy from photovoltaic cells. Although ethanol is a low carbon vector as a transition path for a decarbonized society, there is a necessity to understand that it maintains only 1% of the solar exergy, unlike photovoltaic cells that can reach efficiency as high as 20% for producing electrical energy (or exergy).

9.2 Household Performance and Daily Habits

This section utilizes the concept of Life-Cycle Assessment (LCA), a widely used method to complement thermodynamic assessments. LCA enables the comprehensive mapping of potential environmental impacts associated with a process, product, or activity. It allows for the quantification of consumption habits' implications across various contexts, such as transportation [22], waste management [23, 24], and food production systems [25–27]. Taking a further step, the concept of "extended exergy analysis" [28] combines exergy and LCA into a unified approach. LCA comprises four distinct steps [29]: goal and scope definition, inventory analysis, impact assessment, and interpretation. In the first step, we establish the purpose and boundaries of the LCA. The functional unit, to which all inputs and outputs relate, is also defined. The second step accounts for the material and energy flows associated with the defined functional unit. The third step focuses on selecting a method for assessing environmental impacts. Finally, results are obtained and interpreted, and it is possible to conclude with the proper indicators.

Ensuring acceptable and rational use of exergy is a significant challenge for society within the coming years, as it is crucial to guarantee that future generations can access these resources with adequate quality while minimizing environmental impacts. However, the current generation must also address the consequences of increased CO_2 levels in the atmosphere. In 2020, Brazil's GHG emissions per capita associated with the energy mix amounted to 1.87 ton of $CO_{2,eq}$/hab, with an annual growth rate of 1.4% [30]. These emissions are closely linked to the proportion of renewable sources, particularly hydroelectric power, in Brazil's electricity mix. However, the amount of non-renewable energy sources is increasing in absolute numbers. In 2020, renewable energy sources in electricity generation accounted for 84.8% in Brazil, but their share in the overall domestic energy supply (energy matrix) was 48.4%. The proportion of non-renewable energy sources in electricity generation reflects the influence of climate conditions (rainfall regime) and energy demand, as most dispatchable thermal power plants serve as a backup during periods of lower rainfall. Only one thermal power plant operates based on solar radiation in Rosana, São Paulo.

Despite the increasing investments in renewable sources like solar and wind, the share of renewable energy sources in the energy mix in 2020 was 48.4%, which was only slightly higher than the figure in 2011 (43.6%) [31] the lowest point was reached in 2014 due to drought conditions: 39.5%. In 2020, the Brazilian residential sector accounted for 10.8% of the country's final energy consumption. The primary energy source is electricity, followed by firewood. However, natural gas consumption showed the highest increase compared to 2019 [32]. In terms of overall energy consumption in the residential sector, renewables contribute to 67% [30]. It is essential to invest in energy efficiency and conservation to decrease the environmental burdens associated with energy generation and consumption. In Brazil, significant initiatives targeting consumers are led by INMETRO-PROCEL and PETROBRAS-CONPET [33]. These programs implemented various actions with emphasis on the National Energy Conservation Label, which certifies products and appliances based on their energy efficiency rates available to the public. For a more in-depth analysis of these programs, please refer to [34] and the earlier sections of this text.

9.3 Methods

In this study, a control volume is defined, encompassing a representative Brazilian household, to analyze the inputs and outputs necessary to meet daily needs. For each input and output, both the embedded exergy (in MJ) and GHG emissions (in $kgCO_{2,eq}$) are quantified. The embedded exergy for food reflects the cumulative energy used during the production process, while for fuels, electricity, and waste, it represents the actual exergy content of the respective flows (encompassing both physical and chemical exergy components). The control volume employed in this analysis is illustrated in Fig. 9.2.

A sensitivity analysis is performed using the Engineering Equation Solver [35] to assess the exergy efficiency of the energy conversion processes within the control volume. The exergy efficiency of the household is calculated by considering the efficiency of each individual process and its contribution to the overall energy consumption, following [36].

9.3.1 Residential Appliances

This subsection focuses on the electricity consumption of the air conditioner, electric shower, and refrigerator. The stove is assessed as a non-electrical appliance used to heat water. Wall simulations used manufacturer data and information provided by INMETRO [37]. This section is based on [11, 34] where the model of the

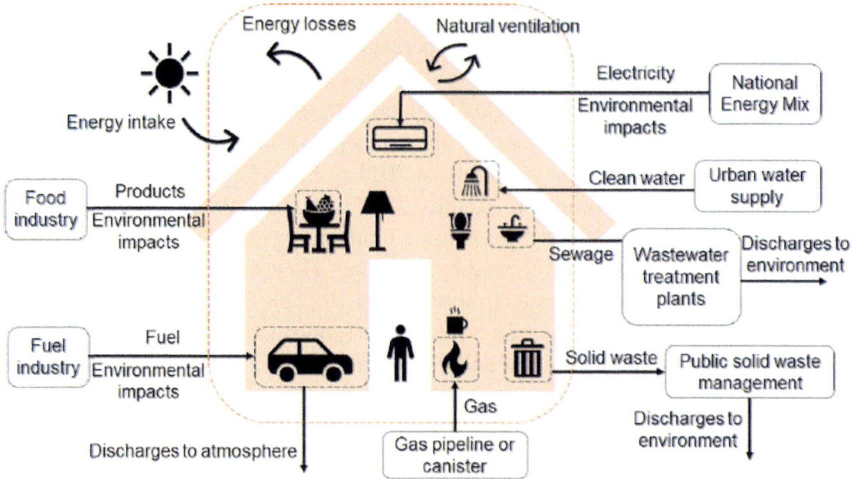

Fig. 9.2 Proposed control volume with its inputs and outputs, and respective suppliers. *Source* [11]

air conditioning system is proposed. In addition to the consumption and efficiency data obtained from manufacturers and INMETRO, the simulations incorporate specific regulations and ordinances published by INMETRO that outline the test methods [38, 39]. The air conditioning usage in the simulations is assumed to be 8 hours per day. The compressor operates 75% of the time [40] to maintain a temperature of 23 °C (a conservative value), and no air conditioning is required when temperatures are close to the thermal comfort range. For modeling purposes, the fluid inlet temperature in the evaporator is considered constant at 10 °C. The fluids used in the refrigerator and the air conditioner simulations are R-134a and R-410a, respectively. The evaporator unit of the refrigerator is located inside the freezer compartment, with a constant temperature of $-6\ ^\circ\mathrm{C}$.

The electric shower and gas heater were modeled using the method proposed in Sect. 2.3.1. The gas stoves are fueled by natural gas (NG) or liquefied petroleum gas (LPG) and are represented in the model by a stove and general water heaters that operate on natural gas. The control volume for this appliance is depicted in Fig. 9.3.

Fig. 9.3 Representation of a stove and general water heaters that operate on natural gas. *Source* [11]

Equation (9.1) is employed for the stove. The simulation assumes the heating of 0.5 kg of water from 25 °C to 100 °C. The typical energy efficiency is 63% for stove burners (based on data obtained from the Brazilian Labeling Program for grade-A stoves [37]).

$$\eta_{ex} = \frac{\dot{m}_{water}\Delta b_{water}}{\dot{m}_{NG} b_{NG}} \tag{9.1}$$

where $\dot{m}_{NG} b_{NG}$ is the chemical exergy of natural gas and $\dot{m}_{water}\Delta b_{water}$ is the increase in the exergy of water.

9.3.2 Supply Chains of Food, Electricity, and Fuel

The supply chains of electricity, food, fuel, and waste are depicted in a summarized way in Figs. 9.4, 9.5, 9.6, 9.7, and 9.8. However, it is crucial to highlight the importance of extending the analyses from cradle to grave, capturing the full life cycle of a process and accurately quantifying the irreversibilities and carbon footprint associated with it.

9.3.3 Agri-Food Systems

Figure 9.4 shows the energy conversion process from the beginning of the plantation to its crop processing.

To account for the variability in environmental impacts of agri-food systems, this study compares two types of diet: an animal protein-based diet (referred to as standard) and a vegetarian diet, following the approach of [41]. The study considers different food groups and their corresponding amounts within the food pyramids, as outlined by Philippi [42], and Venti and Johnston [43] for both types of diet.

This comparison serves as an initial step to evaluate the energy and carbon intensity differences between the two diets, with the vegetarian one generally considered to have lower energy intensity in the production stage. However, the requirement

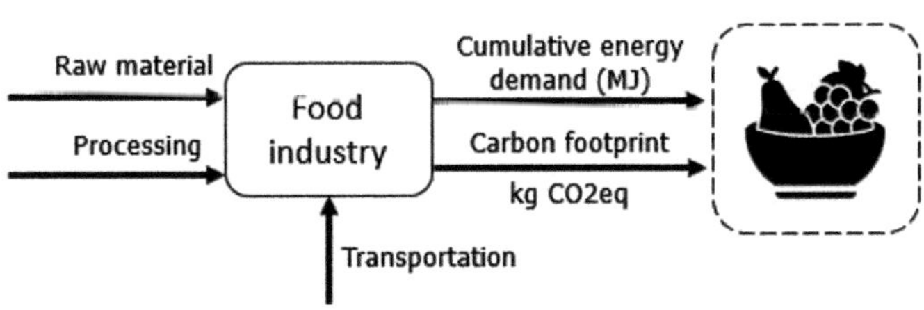

Fig. 9.4 Simplified relationship between agri-food businesses and the supplied products. *Source* [11]

for implementing certain restrictions and the need for public policies to label products with their origin (country, state, city) are evident, as discussed by [11]. The production chains of agri-food systems comprise stages such as clearing land for cultivation or livestock and employing raw materials and infrastructure for operation (e.g., machinery and buildings).

As the production system grows and develops, additional demands arise, including animal feed, fertilizers, packaging materials, and utilities (such as water, electrical energy, and fuel). It is worth noting that the sources of these utilities can be renewable or non-renewable, with the former being the most usual. Once the product is ready for the market, it undergoes transportation to distribution centers and is purchased by consumers. The journey of products from production to the consumer's home encompasses a series of processes with embedded energy and GHG emissions.

The path taken by these products, whether by vehicle or walking mode, can vary significantly and involves numerous variables. These processes vary significantly among the different types of food considered.

- different rates and processes of land-use change, as well as varying rates of water, electrical energy, and transportation;
- employ different technologies in irrigation, packaging, processing, and other areas;
- different energy conversion processes are required to achieve the final product, including industrial-scale production, family farming, or other farming methods.

Gokbulak [44] has explored potential gains in these processes, such as the fact that 1 ton of fast food with correct drying could generate approximately 3.5 GW. Additionally, Nazir and Ozilgen [45] identified potential reductions in the food chain production to reduce $CO_{2,eq}$ emissions in the productive chain of bovine meat. Validated LCA studies were employed to model the production chains of common food items consumed in Brazil, considering both standard and vegetarian diets. The results of these studies were compiled in Table 9.1.

The carbon footprint of each group is discussed in detail in [11]. It is worth noting that Brazilian food habits, as highlighted by Carvalho et al. [59], have been remarked upon, with the population of São Paulo state consuming amounts of meat higher than the average (surpassing the recommended 500 g per week). On the other hand, Monteiro et al. [60] found that only one in eight interviewees reported eating the recommended daily intake of vegetables and fruits in Brazil.

Philippi et al. [42] and Venti-Johnston [43] proposed food pyramids, which are employed as a reference of a correct ingestion and the nutritional values of types of food. The selected foods in this study represent common food groups in the daily life of a middle-class citizen (who does not have any food shortages). These include beef, chicken, milk, eggs, white rice, soy (despite the more significant part of the national soy being exported), beans, bananas, carrots, lettuce, and tomatoes, among others. The quantities of each food used in the analysis are specified in Table 9.2 for the standard diet and Table 9.3 for the vegetarian diet. It is important to emphasize that there is no value judgment imposed on the dietary choices, but rather the comparison of two viable scenarios based on quantitative results derived from

Table 9.1 References that were used to obtain data on GHG emissions and accumulated energy demand (DAE) for food. The first column shows the food groups, followed by the type of food in the second column, the literature reference for DAE and GHG

Group	Food	Ref. DAE and Ref. GEE
Meat	Bovine	[46]
Meat	Chicken	[47]
Dairy	Cow milk	[48]
Grains	White rice	[49] and [26]
Grains	Soybean	[50] and [25]
Grains	Beans	[51]
Fruits	Bananas	[52]
Vegs.	Carrot	[53]
Vegs.	Lettuce	[27] and [54]
Vegs.	Tomatoes	[53]
Carb.	Bread	[55]
Carb	Pasta	[56]
Sweet	Sugar	[57]
Oils	Olive oil	[58]

Table 9.2 The amount of food in grams consumed daily–standard diet, calculated based on [42]. Where in the first column it is the group. The second type of food, in the third, the amount in each portion (portion). In the last two columns, the maximum and minimum consumption

Group	Food	Portion (g)	Max. (g)	Min. (g)
Meat	Bovine meat	100	100	50
Meat	Chicken	75	75	37,5
Dairy products	Cow milk	182	546	546
Grains	White rice	124	372	206,67
Grains	Soybean	12,33	6,17	6,17
Grains	Beans	86	43	43
Fruits	Banana	43	215	129
Vegetables	Carrot	34	56,67	45,33
Vegetables	Lettuce	120	200	160
Vegetables	Tomatoes	109	181,67	145,33
Carbohydrates	Bread	50	150	83,33
Carbohydrates	Pasta	120	360	200
Sweet	Sugar	28	56	28
Oils	Olive oil	8	16	8

Source [42]

Table 9.3 The amount of food in grams consumed daily—standard diet, calculated based on [43]. Where in the first column it is the group. The second type of food, in the third, the amount in each portion (Portion). In the last two columns, the maximum and minimum consumption. Based on: [43]

Group	Food	Portion (g)	Max. (g)	Min. (g)
Dairy products	Cow milk	182	546	546
Grains	White rice	124	413	248
Grains	Soybean	12,33	18,5	12,33
Grains	Beans	86	129	86
Fruits	Banana	75	150	75
Vegetables	Carrot	34	68	34
Vegetables	Lettuce	120	360	240
Vegetables	Tomatoes	109	218	109
Carbohydrates	Bread	50	166,67	100
Carbohydrates	Pasta	120	400	240
Sweets	Sugar	0	0	0
Oils	Olive oil	2,67	8,01	5,34

thermodynamic analysis. The quantities were calculated based on the daily calorie levels recommended for each food group and the grams per serving provided by [42] and [43]. The goal was to create a diverse diet that includes multiple types of food within each food group.

9.3.3.1 Electric Grid and the Electrical Energy Supplied to Consumers

In the residential sector, the primary energy source is electrical energy, which is used for general appliances and lighting, followed by wood, mainly for cooking purposes. Wood has been progressively shifted towards the use of liquefied petroleum gas (LPG) and natural gas (NG) for cooking and sometimes water heating [32]. In terms of electrical energy generation and supply, the Brazilian electricity mix strongly relies on hydropower, supplemented by fossil-fueled thermal power plants (primarily natural gas, coal, and oil), leading to a mix considered for the analyses is 64.9% hydropower, 9.3% natural gas, 3.3% coal, 2.5% nuclear, 8.4% biomass, 2% oil, 1% solar, and 8.6% wind [30]. The associated GHG emissions for the 2019 Brazilian electricity mix were 104.1 kg$CO_{2,eq}$/MWh [32]. Figure 9.5 provides a simplified representation of energy use and the supply of this vector to consumers.

Electrical energy plays an important role in the exergy balance of society and is a vector that demands high energy sources in the production chain [10]. Electrical energy is required by a variety of electrical appliances, including televisions, laptops, coffee machines, washing machines, refrigerators, electric showers, and air conditioners. The average usage of the remaining electrical appliances was based

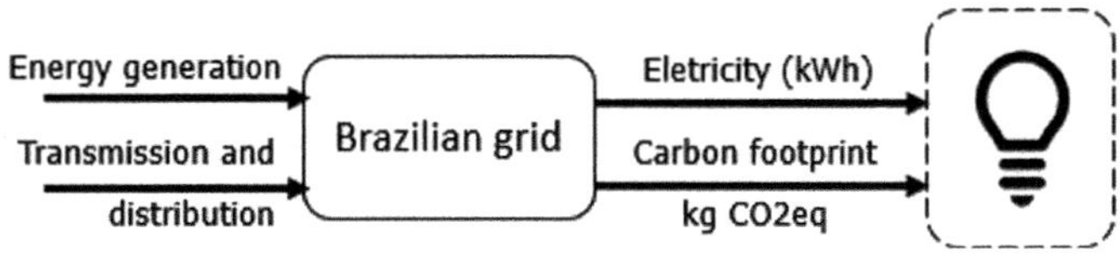

Fig. 9.5 Simplified relationship between the national electric grid and the electricity supplied to consumers. *Source* [11]

on manufacturer data and information from INMETRO [37]. Consumption habits are aligned with following PROCEL guidelines [33]. It is assumed that the electric grid supplies 100% of the electrical energy demanded, with no on-site generation. As illustrated in Fig. 9.5, the grid operates as an interconnected system that receives electricity generated from various power plants and redistributes it throughout the connected subsystems via transmission lines. Local distributors supply it to the end consumers.

9.3.4 Demanded Fuels for Transportation

Figure 9.6 provides an overview of the energy use in the transportation sector, specifically focusing on passenger vehicle applications. First, it is common sense that modal changes, such as increased use of public transportation, walking, cycling, and other modes, are crucial. Nevertheless, these are not always faster, safer, or more comfortable, making this scenario with unrealistic projections to change in the near future.

The fuels used for transportation are represented by their lower heating value (LHV) and are adapted for the exergy basis. The analysis follows a well-to-wheel approach to quantify GHG emissions. The LHV and GHG emissions for each fuel are presented in Table 9.4. Daily transportation patterns reflect typical middle-class practices, with road transportation in a private passenger car [68] in the city of São Paulo. These data are different from the ones in [16, 18].

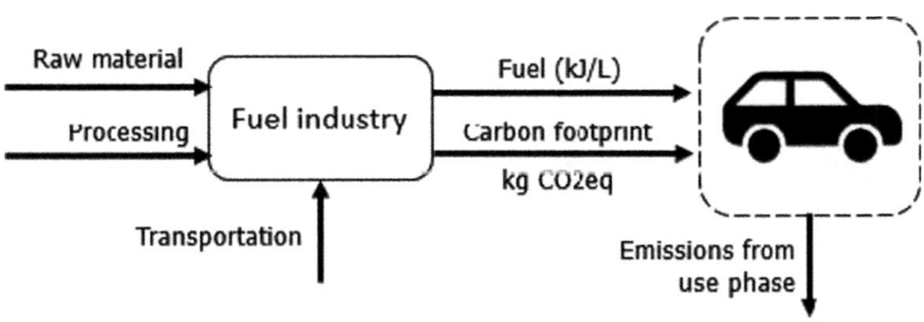

Fig. 9.6 Simplified relationship between the demanded fuels for transportation and its respective industry. *Source* [11]

Table 9.4 Lower heating value (LHV) and greenhouse gas emissions (GHG), from well-to-wheel for selected fuels. Data obtained in [61]

Fuel	LHV (MJ/L)	GHG emissions ($gCO_{2,eq}$/km)
Gasoline	28.9	159.83
Ethanol	21.3	31.53
Diesel	35.6	274.59

9.3.5 Solid Residues Produced and the Management Services

Regarding waste management, the chemical exergy of solids and liquids is taken into account, as it represents the environmental changes caused by these waste materials. In the case of water, it is assumed that 100% of the daily water demand is discarded and then directed to wastewater treatment plants [62, 63], as shown in Fig. 9.7. Figure 9.8 illustrates the relationship between solid waste and waste management.

In Brazil, solid waste is typically not sorted [64], which means that solid waste in Brazil does not usually have a proper destination [65]. Waste is a contributor to greenhouse gas emissions [66], with primary sources including methane emissions from landfills, methane emissions from sewage, and nitrous oxide emissions according to IPCC (Intergovernmental Panel on Climate Change). Methane gas is of particular concern due to its significantly higher global warming potential compared to carbon

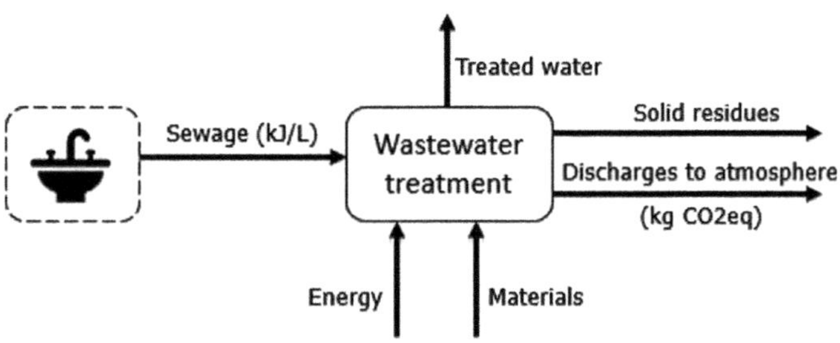

Fig. 9.7 Simplified relationship between the liquid residues produced and the management services. *Source* [11]

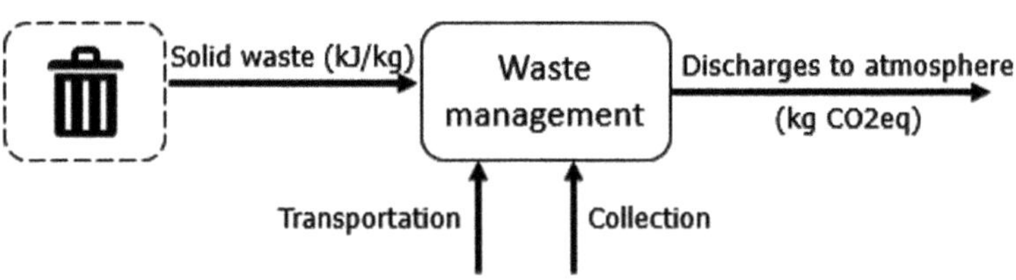

Fig. 9.8 Simplified relationship between the solid residues produced and the management services. *Source* [11]

dioxide over a 100-year period, with a factor of 25 times higher, and waste generation will increase by 60% by 2050 [67], further emphasizing the critical nature of this issue. Furthermore, it is important to highlight that the emission of methane into the atmosphere represents the loss of a valuable exergy resource that could be utilized as a potential fuel for other processes. The exergy assessment in this study utilized data from [23, 69]. For the GHG emissions, [24, 70] were followed.

Concerning Fig. 9.8, the treatment of organic solid wastes is a topic of increasing attention in the literature, as evidenced by a review of Brazilian literature highlighting the importance of digestion, incineration, and pyrolysis in improving the energy generation (biogas and electrical energy from solid wastes), production of pure CH_4 and pure CO_2 and other fuels with high market values. Moreover, the digestate can be used as fertilizer (rich in NPK) and is suitable for (i) urban, (ii) family farms, and (iii) sugarcane production, decreasing the dependence on mineral fertilizers with a high content of carbon footprint [71].

9.4 Results and Discussion

Considering the activities of daily life described earlier, Fig. 9.9 presents the results of greenhouse gas (GHG) emissions associated with diet, electrical energy consumption, fuels, and waste. Two scenarios are compared: maximum emissions, represented by the consumption of gasoline and a standard diet, and minimum emissions, represented by the consumption of ethanol fuel and a vegetarian diet. Note that despite the low emissions of ethanol, its production has lower exergy efficiency compared to electrical energy production by photovoltaic cells [20, 21]. Moreover, an internal combustion engine has lower efficiency than an electrical one. Nevertheless, its carbon footprint is comparable when considering the well-to-wheel analysis of the fuel [17]. However, upon examining the life-cycle analysis of the vehicle, it reveals a higher carbon footprint for electric vehicles (EVs, PHEVs) due to the production of batteries [16, 18], which increases the market share of HEVs over ethanol/gasoline vehicles (flex-power) in countries such as Brazil [18].

From Fig. 9.9, it is observed that on an annual basis, the GHG emissions can be 5,209 kg $CO_{2,eq}$/year for the maximum emission scenario (realistic), or as low as 1,824 kg $CO_{2,eq}$/year considering a vegetarian diet and use of ethanol fuel (a more optimistic scenario in terms of fuel consumption for transportation). However, Brazil is implementing policies aimed at reducing the environmental impact of light-duty vehicles, although with less emphasis on electrification guidelines. In this context, ethanol fuel presents an opportunity for achieving faster gains, leveraging existing infrastructure. The premise of using alternative modes, such as walking, cycling, and public transportation, is essential for effective decarbonization. However, significant improvements in security and attractiveness are needed to encourage greater adoption by consumers.

Therefore, the internal combustion engine is expected to remain the dominant power source in the vehicle fleet for some time. The transportation sector remains

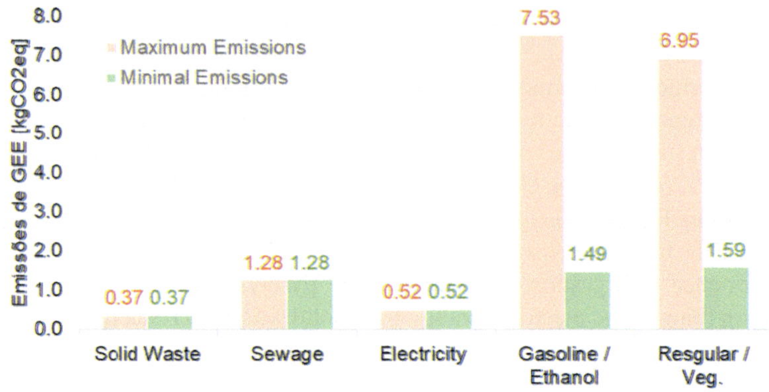

Fig. 9.9 Maximum and minimum daily GHG emissions calculated, using two different scenarios for fuels and food consumption, in $kgCO_{2,eq}$. *Source* [11]

one of the most significant sources of emissions in Brazil, significantly increasing the carbon footprint of products sold in cities, as heavy vehicles with diesel engines are the primary form of transportation, with limited train lines. Figure 9.9 illustrates the potential of ethanol as an intermediate solution to mitigate emissions in light-duty transportation in Brazil, towards a decarbonized modal system. Electric cars still represent a tiny percentage of the Brazilian fleet [19], primarily due to their high prices. The main conclusion is that Brazil has time to shift its powertrain, and efforts must be made in the transportation sector to ensure products flow with higher efficiency and lower carbon emissions than solely transporting them using trucks fueled with diesel.

Figure 9.9 shows that the GHG emissions associated with ethanol correspond to only 20% of gasoline emissions [19]. Mady et al. [34] have pointed out that the National Energy Conservation Label considers ethanol to have zero emissions (probably only considering operation-related emissions). Either way, replacing gasoline with ethanol is already capable of contributing significantly to decarbonization, especially considering the population's generalization of incentives and the widespread adoption of biofuels.

Dietary habits also play a significant role in GHG emissions. The vegetarian diet considered here was found to contribute only 34% of the GHG emissions compared to the standard diet. Although a personal habit and preference, it encompasses an entire production chain, with significant impacts associated with the origin of the protein consumed as discussed in previous studies [72–74]. Global food systems are responsible for approximately 30% of total anthropogenic GHG emissions, and shifting towards plant-based diets and local food consumption can help mitigate these emissions [75]. These findings emphasize the potential of individual dietary changes in contributing to climate change mitigation, and the emergence of semi-vegetarian and light semi-vegetarian diets reflects growing concerns over animal welfare and ecological considerations [76]. These changes are desirable, especially if they can be implemented on a broader scale. The Meatless Mondays campaign, aimed at reducing meat consumption by 15%, has shown promising results. For instance, in a large urban US school district, the entrees served during Meatless Mondays were associated with −74% GHG emissions [77]. It is important to note that the intention is not to impose a specific diet, but rather to test and explore each example, as dietary choices are influenced by various factors such as personal taste, culture, lifestyle, and external factors like public policies, food prices, and food availability [78]. However, it is important to note that adopting a vegetarian-based diet does not automatically guarantee environmental friendliness: Pereira and Mady [79] verified that soy cultivation could impact land use as new areas are cleared (deforestation). Providing such information to consumers is crucial, as already occurs in the European Union [80]. This is an opportunity for the consumer to achieve a better understanding and context of the sustainability debate and make more informed choices when possible, as discussed by [81]. The conclusion is that some changes can lead to an increase in the carbon footprint on a daily routine and, even more, can lead to campaigns or policies that are greenwashing.

Analogously to the results presented for GHG emissions, the exergy flows are shown in Fig. 9.10. On a daily basis, each individual has an average exergy consumption that corresponds to 52437 MJ annually. It should be noted that the only change that may affect the exergy demand between these two scenarios is the different modeled diets. However, the removal of animal protein from the diet and the adjustment of ingested portion sizes did not result in a significant impact, as Brazilian livestock has relatively low energy intensity. In the Brazilian household model considered, personal transportation was found to have the highest energy and exergy intensities. If public transportation was considered in the analysis, it is likely that these values would decrease, as the demand would be shared among all passengers. Public transportation has been evaluated as a strategy to conserve energy and reduce emissions in Canada [82]. One crucial observation is that a person needs the same

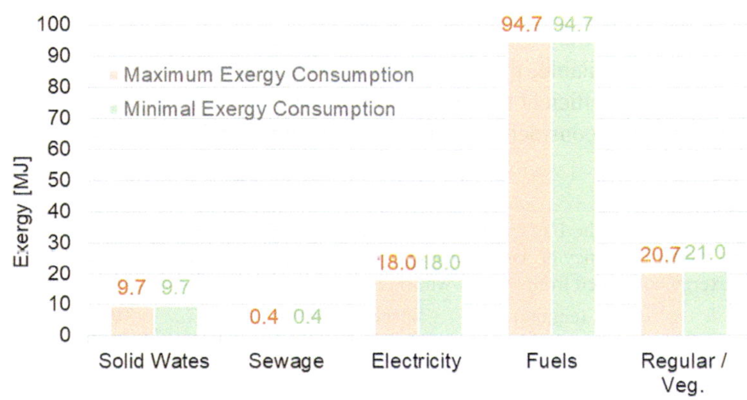

Fig. 9.10 Minimum and maximum exergy calculated for each flow, using two different scenarios for fuels and food consumption. *Source* [11]

amount of exergy to survive. Therefore, the energy and exergy consumption should be similar. Nevertheless, based on the food pyramid, it is unexpected that both dietary paths led to similar exergy production "costs."

It is interesting to compare Figs. 9.9 and 9.10 due to their complementary nature. From the perspective of exergy analysis, there appears to be an equilibrium in the supply and expenditure of exergy to meet the average daily necessities of the household. However, the $CO_{2,eq}$ emissions show a significant difference based on personal choices, highlighting the carbon footprint associated with each path. Therefore, both are complementary analyses. There are some indexes such as renewability [84].

Using the results of [17] for a model using internal combustion engine vehicles (ICEV) and battery electric vehicles (BEV). Vehicle power sources in Brazil are characterized by fuel blends as defined by INMETRO [83] and ANP (Brazilian National Agency for Petroleum, Natural Gas and Biofuels), where gasoline available for light vehicles is blended with ethanol (E22, up to 27% anhydrous ethanol by volume, or close to 24% by mass), which is also available in its pure form, mixed with a small percentage of water (E100, hydrous ethanol). The anhydrous ethanol content in E22 gasoline can vary from 18% to 27.5% by volume. It is important to highlight that, differently from the electric mix, the energy mix has a high percentage of non-renewable energy sources [11]. The results of Feliciano et al. [17] show a clear distinction of the exergy consumed and efficiency of electric vehicles in comparison with the traditional internal combustion engines on E100 or E22 (Table 9.5).

Nonetheless, when changing the electrical matrix from the Brazilian to the average electric mix in Europe, shown in Feliciano et al. [17], it is also possible to realize that the exergy input and destroyed exergy over the cycle for electric vehicles might increase due to the high usage of combustion-based electric power generation (usually fossil), which is an indication of how exergy destruction may vary as a function of the electrical mix for the location where the vehicle is charged, as seen in Table 9.6. Moreover, this fact is significant to Brazil, where there are populous and urban areas where energy production varies seasonally. When electricity is in peak demand, there is an increase in the electric power offered by thermoelectric stations. These particularities may decrease the advantages of BEVs over ICEVs. Moreover, when ethanol is available, there are even more variables to assess since it is less energy- and exergy-efficient than BEVs, but with zero CO_2 from tank to wheel. This information must be considered in the creation of public policies to decrease taxes on

Table 9.5 Results for the FTP75 cycle. Where the first column it is indicated the type of fuel, flowed by exergy efficiency of the engine (η_b), performed power, exergy efficiency of the vehicle (η_b) and exergy consumed from well-to-wheel ($B_{consumed}$)

Source	η_b (engine) (%)	W (performed power) (MJ)	η_b (%)	$B_{consumed}$ (WtW) (MJ)
E22	20.9	9.8	19.5	49.9
E100	21.8	9.8	19.4	50.1
Electricity	97.8	11.6	50.3	25.0

Table 9.6 Results for the electric vehicle with the European electric mix

Cycle	Electric matrix exergy efficiency (%)	Well-to-wheel exergy efficiency (%)	Exergy consumed (MJ)
City	36.65	35.80	35.14
Highway	36.65	36.10	35.40

Table 9.7 CO_2 emissions for the different urban cycles and fuels in Brazil and for electric cars in Europe. It was not considered the emissions of battery production solely the vector

Cycle	Fuel	CO_2—Brazil (g)	CO_2—Europe
Urban	Electricity	321.96	1094.65 g
Urban	E22	3599.49	–
Urban	E100	1026.75	–

electric vehicles over ethanol, for instance. This fact is more prominent when there is a change in the Brazilian tariff, named the "red tariff," where more energy is produced with fossil fuels, different from other seasons, when energy from renewable sources is enough to cover demand.

Equivalent CO_2 emissions can be found in Table 9.7, considering specific CO_2 emissions for the Brazilian energy matrix, E100, and E22 fuel, as stated in [17]. Note that the effect of renewability is not considered in exergy efficiency. There are authors in the literature that discuss this matter, although there is still no consensus on the best indicator [84]. The most important result is that ethanol significantly impacts the decarbonization of the transport sector. It always has lower global warming potential, in equivalent CO_2 than electric cars in Europe. This result considers the CO_2 emissions in the whole production chain and only the fuel itself or electricity generation, without the battery production and discharge. Similar results may be found in winter, where the electric mix in regions such as São Paulo uses more thermoelectric electric energy generation with non-renewable fuels. This raises the following question for policymakers: in these seasons, is it better to fill the tank with ethanol or charge it with electricity?

9.5 Concluding Remarks

In this text, Brazilian consumption habits are represented by an infinitesimal element in the form of an individual, aiming to analyze how these relate to concepts of sustainable development and climate change mitigation. The aim was to evaluate essential demands for the daily life of this individual, and the following were selected: electrical energy, fuels for transportation, food, basic sanitation, and solid waste management. This chapter is helpful for a variety of audiences interested in understanding the cause-and-effect relationship between consumption habits during daily routines, the cycle of product production and usage, and the problem of climate change and energy intensity. The purpose is to reach academia, decision-makers, and policymakers. For the latter, the usefulness of the work done in proposing policies and interventions, as well as the analyses surrounding this axis, includes the identification of trends, synergies, and trade-offs, and the experiences from sectors where its application is relevant.

A middle-class household (control volume) with a single individual is proposed, where the inputs and outputs represent their daily needs through the lens of exergy analysis and GHG emissions (life-cycle analysis). Furthermore, a sensitivity analysis was performed using the exergy efficiency indicator to compare the equipment used in daily energy conversion processes. The tests were conducted for two scenarios of potential GHG emissions, varying the fuel used in individual transportation and the dietary choices made. For the base scenario, the results of a daily potential of 16.64 and an annual potential of 6,075 $kgCO_{2_{eq}}$, while the alternative scenario has a daily potential of 5.24 and an annual potential of 1,913 $kgCO_{2_{eq}}$.

The alternative scenario (vegetarian and ethanol) corresponds to approximately 30% of the baseline scenario, thus demonstrating great potential for improvement at the individual level through changes in consumption habits associated with fuels, including the replacement of gasoline with ethanol, and in diet, with the restriction of beef consumption. Furthermore, these conclusions highlight these items as urgent points for mitigating emissions at the sectoral and national scale due to the intensity of consumption and adverse conditions during production and/or use.

In terms of exergy, the results obtained were an average total demand of 52.437 MJ per year and 143.66 MJ per day. There was no significant variation between the baseline and alternative scenarios, as the exergy demand is similar. Therefore, there is no noticeable change in daily energy expenditure. Future works may refer to the exergy efficiency of the two different scenarios, which is an object of future work.

References

1. Reistad GM (1980) Available-energy utilization in the United States. Thermodynamics: Second Law Analysis. ACS Symposium Series Vol. 122 Chapter 6, 93–109
2. Wall G (1986) Exergy: a useful concept (p. 328). Chalmers Tekniska Hogskola
3. Wall G (1990) Exergy conversion in the Japanese society. Energy 15(5):435–444
4. Biondi A, Sciubba E (2021) Extended exergy analysis (EEA) of Italy, 2013–2017. Energies 14(10):2767
5. Utlu Z, Hepbasli A (2005) Analysis of energy and exergy use of the Turkish residential-commercial sector. Build Environ 40(5):641–655
6. Ertesvag IS, Mielnik M (2000) Exergy analysis of the Norwegian society. Energy 25(10):957–973
7. Ertesvåg IS (2001) Society exergy analysis: a comparison of different societies. Energy 26(3):253–270
8. Schaeffer R, Wirtshafter RM (1992) An exergy analysis of the Brazilian economy: from energy production to final energy use. Energy 17(9):841–855
9. Rosen MA (1992) Evaluation of energy utilization efficiency in Canada using energy and exergy analyses. Energy 17(4):339–350

10. Mosquim RF, Oliveira-Junior S, Mady CEK (2018) Modelling the exergy behavior of São Paulo State in Brazil. J Clean Prod 197:643–655
11. Pereira MTRM, Carvalho M, Mady CEK (2022) Adressing energy demand and climate change through the second law of thermodynamics and LCA towards a rational use of energy in Brazilian households. Entropy 24(11):1524
12. Jin P, Meng Z, Yan K, Chen B, Zhang B (2023) Ecological accounting of the Chinese society 2012–2020 based on extended exergy. J Clean Prod 417:137929
13. Hossain S, Chowdhury H, Chowdhury T, Ahamed JU, Saidur R, Sait SM, Rosen MA (2020) Energy, exergy and sustainability analyses of Bangladesh's power generation sector. Energy Rep 6:868–878
14. Mady CEK, Bresciani AE, Paiva JL, de Jardin-Junior RN, Guardani R, Oliveira-Junior S (2018) Thermodynamic assessment of MVR implementation in multistage evaporation plants. J Brazilian Soc Mech Sci Eng 40(8):395
15. Lima LS, Mady CEK (2021) Energy and exergy analysis of an absorption and mechanical system for a dehumidification unit in a gelatin factory. Entropy 23(4):415
16. Rovai FF, Cal Seixas SR, Mady CEK (2023) Regional energy policies for electrifying car fleets. Energy 278:127908
17. Feliciano HNF, Rovai FF, Mady CEK (2023) Energy, exergy, and emissions analyses of internal combustion engines and battery electric vehicles for the brazilian energy mix. Energies 16(17):6320
18. Rovai FF, Mady CEK (2025) Regional environmental comparison of electrification and ethanol blends in light vehicles. Clean Energy Syst 11:100189
19. Mosquim RF, Mady CEK (2022) Performance and efficiency trade-offs in Brazilian passenger vehicle fleet. Energies 15(15):5416
20. Righetto FG, Mady CEK (2023) Exergy Analysis of a Sugarcane Crop: A Planting-to-Harvest Approach. Sustainability 15(20):14686
21. Righetto FG, Mady CEK (2024) A Thermodynamic Comparison of the Exergy Production from Sugarcane and Photovoltaic Modules in Brazilian Energy Transition Context. Energies 17(19):4940
22. Vargas JEV, Falco DG, da Silva Walter AC, Cavaliero CKN (2019) Seabra J E A Life cycle assessment of electric vehicles and buses in Brazil: effects of local manufacturing, mass reduction, and energy consumption evolution. Int J Life Cycle Assess 24:1878–1897
23. Jadhao SB, Shingade SG, Pandit AB (2017) Bakshi B R Bury, burn, or gasify: Assessing municipal solid waste management options in Indian megacities by exergy analysis. Clean Technol Environ Policy 19:1403–1412
24. Coelho LMG (2018) Lange L C Applying life cycle assessment to support environmentally sustainable waste management strategies in Brazil. Resour Conserv Recycl 128:438–450
25. Castanheira ÉG, Freire F (2013) Greenhouse gas assessment of soybean production: implications of land use change and different cultivation systems. J Clean Prod 54:49–60
26. Nunes FA, Seferin M, Maciel VG, Flôres SH, Ayub MAZ (2016) Life cycle greenhouse gas emissions from rice production systems in Brazil: A comparison between minimal tillage and organic farming. J Clean Prod 139:799–809
27. Bartzas G, Zaharaki D, Komnitsas K (2015) Life cycle assessment of open field and greenhouse cultivation of lettuce and barley. Inf Process Agric 2:191–207
28. Sciubba E (2001) Beyond thermoeconomics? The concept of extended exergy accounting and its application to the analysis and design of thermal systems. Exergy Int J 1:68–84
29. Guinee J B *Handbook on Life Cycle Assessment: Operational Guide to the ISO Standards*; Springer Science & Business Media: Berlin, Germany, 2002; Volume 7
30. Ministry of Mines and Energy -MME. Brazilian Energy Balance 2020 Year 2019 (In Portuguese) (2020). Available online: https://www.epe.gov.br/sites-pt/publicacoes-dados-abertos/publicacoes/PublicacoesArquivos/publicacao-479/topico-528/BEN2020sp.pdf
31. Ministry of Mines and Energy -MME Brazilian Energy Balance 2022 Year 2021 (In Portuguese) (2022). https://www.epe.gov.br/sites-en/publicacoes-dados-abertos/publicacoes/Paginas/Brazilian-Energy-Balance-2021.aspx

32. Ministry of Mines and Energy -MME. Brazilian Energy Balance 2021 Year 2020 (2021). https://www.epe.gov.br/sites-en/publicacoes-dados-abertos/publicacoes/Paginas/Brazilian-Energy-Balance-2020.aspx
33. INMETRO. ENCE-Etiqueta Nacional de Conservação de Energia: Selo PROCEL de Economia de Energia/ANEXO DA PORTARIA INMETRO No 410/2013. Available online: http://www.inmetro.gov.br/legislacao/
34. Mady CEK, Pinto CR, Pereira MTM (2020) Application of the second law of thermodynamics in brazilian residential appliances towards a rational use of energy. Entropy 22(6):616
35. Klein S, Alvarado F (2022) EES-Engineering Equation Solver, F-Chart Software
36. Pedreira NMS (2017) Cenários de Consumo de Energia Domiciliar No Estado do Rio De Janeiro: A Contribuição da Eficientização Para uma Reflexão Sobre a Pobreza Energética. Ph.D. Thesis, Universidade Federal do Rio de Janeiro, Rio de Janeiro, Brazil
37. National Institute of Meteorology Tabelas de eficiência energética (2022). (In Portuguese) https://www.gov.br/inmetro
38. National Institute of Meteorology. Portaria No 332, de 2 de Agosto de 2021-Aprova os Requisitos de Avaliação da Conformidade para Refrigeradores e Assemelhados-Consolidado (2021). (In Portuguese) http://www.inmetro.gov.br/legislacao/rtac
39. National Institute of Meteorology Portaria no 234, de 29 de Junho de 2020.-Aperfeiçoamento Parcial dos Requisitos de Avaliação da Conformidade para Condicionadores de Ar, estabelecendo o Índice de Desempenho de Resfriamento Sazonal (IDRS), a reclassificação das categorias de eficiência energética e determinando outras providências para a disponibilização destes produtos no mercado nacional (2020). (In Portuguese) https://www.sindhoesg.org.br/dados/publicacoes/pub0003496-b61599b07e6b295cf6cbee47286f2f04.pdf
40. Ministry of Mines and Energy Nota técnica EPE 030/2018: Uso de Ar Condicionado no Setor Residencial Brasileiro: Perspectivas e contribuições para o avanço em eficiência energética (2018). (In Portuguese) https://www.epe.gov.br/pt/publicacoes-dados-abertos/publicacoes/nota-tecnica-epe-030-2018
41. Wallén A, Brandt N (2004) Wennersten R Does the Swedish consumer's choice of food influence greenhouse gas emissions? Environ Sci Policy 7:525–535
42. Philippi ST, Brazilian food pyramid. (2005) Nutr. Today 40:79–83
43. Venti CA, Johnston CS (2002) Modified food guide pyramid for lactovegetarians and vegans. J Nutr 132:1050–1054
44. Gökbulak SK, Nazir S, Tunçel S, Özilgen M (2020) How to benefit from the food waste in the era of global warming? Int J Glob Warm 20:216–236
45. Nazır S, Özilgen M (2020) How to reduce environmental impact of hamburger menu? Int J Glob Warm 22:235–254
46. Cederberg C, Persson U M, Neovius K, Molander S, Clift R (2011) Including carbon emissions from deforestation in the carbon footprint of Brazilian beef
47. da Silva VP, van der Werf HMG, Soares SR, Corson MS (2014) Environmental impacts of French and Brazilian broiler chicken production scenarios: An LCA approach. J Environ Manag 133:222–231
48. Léis C M D Desempenho ambiental de três sistemas de produção de leite no Sul do Brasil pela abordagem da Avaliação do Ciclo de Vida (2013). https://repositorio.ufsc.br/handle/123456789/123127
49. Lin HC (2016) Fukushima Y Rice cultivation methods and their sustainability aspects: Organic and conventional rice production in industrialized tropical monsoon Asia with a dual cropping system. Sustainability 8:529
50. da Silva VP, van der Werf HMG, Spies A (2010) Soares S R Variability in environmental impacts of Brazilian soybean according to crop production and transport scenarios. J Environ Manag 91:1831–1839
51. Carmo HFD et al (2016) Balanço energético e pegada de carbono nos sistemas de produção integrada e convencional de feijão-comum irrigado (In Portuguese). Pesqui. Agropecuária Bras. 51:1069–1077

52. Coltro L, Karaski TU (2019) Environmental indicators of banana production in Brazil: Cavendish and Prata varieties. J Clean Prod 207:363–378
53. Fisher S, Karunanithi A (2014) Contemporary comparative LCA of commercial farming and urban agriculture for selected fresh vegetables consumed in Denver, Colorado. In Proceedings of the LCA Food Conference, San Francisco, CA, USA 8–10:8–10
54. Venkat K (2012) Comparison of twelve organic and conventional farming systems: a life cycle greenhouse gas emissions perspective. J Sustain Agric 36:620–649
55. Tassielli GNB, Monforti-Ferrario F, Renzulli PA (2015) Energy Flows and Greenhouses Gases of EU national breads using LCA approach. Challenges and Opportunities in Using Life Cycle Assessment and Cradle to Cradle for Biodegradable Bio-Based polymers: A Review, 2015, 347–350; Proceedings. International conference on Life Cycle Assessment as reference methodology for assessing supply chains and supporting global sustainability challenges: LCA for Feeding the planet and energy for life
56. Ruini L (May2009) Marino M Lca of semolina dry pasta produced by barilla. In Proceedings of the Sustainable Development: A Challenge for European Research, Brussels, Belgium 26:26–28
57. Seabra JEA, Macedo IC, Chum HL, Faroni CE, Sarto CA (2011) Life cycle assessment of Brazilian sugarcane products: GHG emissions and energy use. Biofuels Bioprod. Biorefining 5:519–532
58. Tsarouhas P, Achillas C, Aidonis D, Folinas D (2015) Maslis V Life Cycle Assessment of olive oil production in Greece. J Clean Prod 93:75–83
59. de Carvalho AM, César CLG, Fisberg RM (2013) Marchioni D M L Excessive meat consumption in Brazil: diet quality and environmental impacts. Public Health Nutr 16:1893–1899
60. Jaime PC (2005) Monteiro C A Fruit and vegetable intake by Brazilian adults, 2003. Cad. Saúde Pública 21:S19–S24
61. Flórez-Orrego D, Silva JAM, Oliveira-Junior S (2015) Exergy and environmental comparison of the end use of vehicle fuels: The Brazilian case. Energy Convers Manag 100:220–231
62. Marinoski AK, Vieira AS, Silva AS (2014) Ghisi E Water end-uses in low-income houses in Southern Brazil. Water 6:1985–1999
63. Barreto D (2000) Residential water profile and internal end uses. Sao Paulo, Brazil, Institute for Technological Researches (IPT)
64. Alfaia RGSM, Costa AM (2017) Campos J C Municipal solid waste in Brazil: A review. Waste Manage Res 35:1195–1209
65. Mollica GJG, Balestieri JAP (2020) Is it worth generating energy with garbage? Defining a carbon tax to encourage waste-to-energy cycles. Appl Therm Eng 173:115195
66. Ritchie H, Rosado P, Roser M (2023) CO2 and greenhouse gas emissions. Our world in data
67. Klausner T et al (2020) Urban greenhouse gas emissions from the Berlin area: A case study using airborne CO2 and CH4 in situ observations in summer 2018. Elem Sci Anth 8:15
68. Menezes E, Maia AG (2017) de Carvalho C S Effectiveness of low-carbon development strategies: Evaluation of policy scenarios for the urban transport sector in a Brazilian megacity. Technol Forecast Soc Chang 114:226–241
69. Mora Bejarano CH (2009) Avaliação Exergoecológica de Processos de Tratamento de Esgoto. Ph.D. Thesis, Universidade de São Paulo, São Paulo, Brazil
70. Lopes TA, Queiroz LM, Torres EA (2020) Kiperstok A Low complexity wastewater treatment process in developing countries: A LCA approach to evaluate environmental gains. Sci Total Environ 720:137593
71. Pinheiro JP, Camiloti PR, Sauer IL, Mady CEK (2025) Exergy analysis of a biogas plant for municipal solid waste treatment and energy cogeneration. Energies 18(11):2804
72. Scarborough P, Appleby PN, Mizdrak A, Briggs AD, Travis RC, Bradbury KE (2014) Key T J Dietary greenhouse gas emissions of meat-eaters, fish-eaters, vegetarians and vegans in the UK. Clim Chang 125:179–192
73. Barthelmie RJ (2022) Impact of dietary meat and animal products on GHG footprints: The UK and the US. Climate 10:43

74. Rippin HL, Cade JE, Berrang-Ford L, Benton TG, Hancock N (2021) Greenwood D C Variations in greenhouse gas emissions of individual diets: Associations between the greenhouse gas emissions and nutrient intake in the United Kingdom. PLoS ONE 16:e0259418
75. Li M, Jia N, Lenzen M, Malik A, Wei L, Jin Y (2022) Raubenheimer D Global food-miles account for nearly 20% of total food-systems emissions. Nat Food 3:445–453
76. De Backer CJ (2014) Hudders L From meatless Mondays to meatless Sundays: Motivations for meat reduction among vegetarians and semi-vegetarians who mildly or significantly reduce their meat intake. Ecol Food Nutr 53:639–657
77. Blondin SA, Cash SB, Griffin TS, Goldberg JP, Economos CD (2022) Meatless Monday National School Meal Program Evaluation: Impact on Nutrition, Cost, and Sustainability. J Hunger Environ Nutr 17:1–13
78. Kapur K (2008) Dunning T A global survey of eating preferences, dietary habits and food beliefs. J Diabetes Nurs 12:109
79. Pereira MTRM, Mady CEK (2021) Exergy analysis as a tool for the rational use of energy. Evaluating everyday resource consumption and CO2 emissions. In: Proceedings of the 34th international conference on efficiency, cost, optimization, simulation and environmental impact of energy systems, Messina, Italy, 28 June–2 July
80. Goudis A (2020) Skuras D Consumers' awareness of the EU's protected designations of origin logo. Br Food J 123:1–18
81. Carvalho M, Santana Freire R, Brito AMVG (2016) Promotion of sustainability by quantifying and reducing the carbon footprint: New practices for organizations. Energy, transportation and global warming. Springer, Berlin, Germany, pp 61–72
82. Litman T (2012) Evaluating public transit as an energy conservation and emission reduction strategy. Victoria Transport Policy Institute, Victoria, BC, Canada
83. INMETRO (2022) The National Institute of Metrology. Consumption tables, energy efficiency of light vehicles, Standardization and Industrial Quality
84. Oliveira-Junior S (2012). Exergy: production, cost and renewability. Springer Science & Business Media

9783032016454